SPOKEN, MULTILINGUAL AND MULTIMODAL DIALOGUE SYSTEMS

SPOKEN, MULTILINGUAL AND MULTIMODAL DIALOGUE SYSTEMS

DEVELOPMENT AND ASSESSMENT

Ramón López-Cózar Delgado

Granada University
Spain

Masahiro Araki

Kyoto Institute of Technology
Japan

John Wiley & Sons, Ltd

Other Wiley Editorial Offices

John Wiley & Sons Inc., 111 River Street, Hoboken, NJ 07030, USA

Jossey-Bass, 989 Market Street, San Francisco, CA 94103-1741, USA

Wiley-VCH Verlag GmbH, Boschstr. 12, D-69469 Weinheim, Germany

John Wiley & Sons Australia Ltd, 42 McDougall Street, Queensland 4064, Australia

John Wiley & Sons (Asia) Pte Ltd, 2 Clementi Loop #02-01, Jin Xing Distripark, Singapore 129809

John Wiley & Sons Canada Ltd, 22 Worcester Road, Etobicoke, Ontario, Canada M9W 1L1

Wiley also publishes its books in a variety of electronic formats. Some content that appears
in print may not be available in electronic books.

British Library Cataloguing in Publication Data

A catalogue record for this book is available from the British Library

ISBN-13 978-0-470-02155-2
ISBN-10 0-470-02155-1

Typeset in 10/12pt Times by Integra Software Services Pvt. Ltd, Pondicherry, India
Printed and bound in Great Britain by Antony Rowe Ltd, Chippenham, Wiltshire
This book is printed on acid-free paper responsibly manufactured from sustainable forestry
in which at least two trees are planted for each one used for paper production.

Contents

Preface

In many situations, the dialogue between two human beings seems to be performed almost effortlessly. However, building a computer program that can converse in such a natural way with a person, on any task and under any environmental conditions, is still a challenge. One reason why is that a large amount of different types of knowledge is involved in human-to-human dialogues, such as phonetic, linguistic, behavioural, cultural codes, as well as concerning the world in which the dialogue partners live. Another reason is the current limitations of the technologies employed to obtain information from the user during the dialogue (speech recognition, face localisation, gaze tracking, lip-reading recognition, handwriting recognition, etc.), most of which are very sensitive to factors such as acoustic noise, vocabulary, accent, lighting conditions, viewpoint, body movement or facial expressions. Therefore, a key challenge is how to set up these systems so that they are as robust as possible against these factors.

Several books have already appeared, concerned with some of the topics addressed in this book, primarily speech processing, since it is the basis for spoken dialogue systems. A huge amount of research papers can be found in the literature on this technology. In recent years, some books have been published on multimodal dialogue systems, some of them as a result of the selection of workshops and conference papers. However, as far as we know, no books have as yet been published providing a coherent and unified treatment of the technologies used to set up spoken, multilingual and multimodal dialogue systems. Therefore, our aim has been to put together all these technologies in a book that is of interest to the academic, research and development communities.

A great effort has been made to condense the basis and current state-of-the-art of the technologies involved, as well as their technological evolution in the past decade. However, due to obvious space limitations, we are aware that some important topics may have not been addressed, and that others may have been addressed quite superficially. We have tried to speak to all the constituencies. The topics cover a wide range of multi-disciplinary issues and draw on several fields of study without requiring too deep an understanding of any area in particular; in fact, the number of mathematical formulae is kept to a minimum. Thus, we think reading this book can be an excellent first step towards more advanced studies.

The book will also be useful for researchers and academics interested in having some reference material showing the current state-of-the-art of dialogue system. It can also be useful for system developers interested in exploiting the emerging technology to develop automated services for commercial applications. In fact, it contains a large number of Internet links where the reader can find detailed information, development Internet sites and development tools download. Professors as well as undergraduate and post-graduate students of Computer Science, Linguistics, Speech and Natural Language Processing, Human-Computer Interaction and Multimodal Interactive Systems will also find this text useful.

Writing this book has been a challenging and fascinating journey down many endless roads, full of huge amounts of information concerned with the technologies addressed. As commented above, given the limitations of space, it has not been easy to come up with a trade-off between discussing the diverse topics in detail, on the one hand, while, on the other, giving a general, wide-range overview of the technologies. We hope the efforts we made to condense this universe of information into just one book will benefit the reader.

Journeying down these roads we have encountered many researchers and companies that have kindly granted permissions to reproduce material in this book. We wish to thank the kind collaboration of them all, and especially the contribution of Jan Alexandersson (DFKI, Germany), Koray Balci (ITC-irst Cognitive and Communication Technologies Division, Italy), Marc Cavazza (University of Teesside, UK), Mark Core (University of Southern California, USA), Jens Edlund (KTH, Sweden), James R. Glass (MIT, USA), Clegg Ivey (Voxeo Corporation, USA), Sanshzar Kettebekov (Advanced Interfaces, Inc., USA), Michael F. McTear (University of Ulster, Northern Ireland), Nick Metianu (IBM Software Group, USA), Yasuhisa Niimi (ATR, Japan), Rainer Stiefelhagen (University of Karlsruhe, TH, Interactive Systems Labs, Germany), Kevin Stone (BeVocal Café, USA), Jan Van Santen (OGI School of Science and Engineering, Oregon Health and Science University (USA), and Yunbiao Xu (Hangzhou University of Commerce, China).

We also would like to thank the AAAI, Elsevier and Springer for their kind permission to reproduce material in this book.

Finally, we would also like to thank very much and in particular the support, help and contribution of the student and scholarship holder, Zoraida Callejas, and Professor Miguel Gea in the Department of Languages and Computer Science at Granada University, Spain.

Ramón López-Cózar Delgado
Granada

Masahiro Araki
Kyoto
April 2005

1

Introduction to Dialogue Systems

1.1 Human-Computer Interaction and Speech Processing

The so-called Human-Computer Interaction (HCI) is a multidisciplinary field in which three main elements are involved: human beings, computers and interaction. Research on HCI is very important because it stimulates the development of new interfaces that reduce the complexity of the interaction and ease the use of computers by non-expert users. In the past, HCI was extremely rigid since the user had to interpret the information provided by the computer expressed in a very different language from the human one. Technological advances have greatly improved the interaction. For example, the first-generation computers that only allowed letters to be displayed have been replaced by multimedia computers that allow reproduction of graphics, videos, sounds, etc., making the interaction much more comfortable. However, classical interaction with computers based on screen, keyboard and mouse, can be carried out only after the user has the minimal knowledge about hardware and software.

An alternative and relatively new way of interacting with computers is based on the processing of the human speech, which allows several advantages in comparison to the classical one based on keyboard, mouse and screen. Among others, speech offers a greater speed for transmitting information, allows other tasks (liberating the user from the need to use his or her hands and/or eyes) to be carried out simultaneously, reveals the identity of the speaker and permits some disabled users to interact with the computer. Speech allows a great expressivity, in fact, human beings express their ideas, feelings, etc. in language. Speech also allows information about the state of mind of the speaker, his/her attitude towards the listener, etc. to be transmitted. Moreover, speech can be transmitted by simple and widely used devices such as fixed and mobile telephones, making possible remote access to a variety of speech-based services.

The start of speech-based interaction with computers can be traced back to 1977, when in the USA several companies started to develop commercial applications at a very low cost, as, for example, the speaking calculator presented by Telesensory Systems Inc. or the Speak'n Spell system by Texas Instruments. Among other applications, this kind of interaction is currently used to interact with program interfaces (e.g. to move the cursor to a

Spoken, Multilingual and Multimodal Dialogue Systems: Development and Assessment   Ramón López-Cózar Delgado
and Masahiro Araki   © 2005 John Wiley & Sons, Ltd

specific position of the screen) and operating systems (e.g. to run programs, open windows, etc.). This type of interaction is also used in dictation systems, allowing the user to write documents without the need to type but only say the words. Speech-based communication is also used to control devices in domestic environments (e.g. turn on lights, ovens, hi-fi sets, etc.) which can enhance the quality of life of disabled people. Moreover, this kind of communication is used to interact with car navigation and other in-car devices, allowing a hands- and eye-free interaction for drivers that increases their safety.

1.2 Spoken Dialogue Systems

Another kind of application of the speech-based interaction is the so-called Spoken Dialogue Systems (SDSs), also called *conversational systems*, that can be defined as computer programs developed to provide specific services to human beings in the same way as if these services were provided by human beings, offering an interaction as natural and comfortable as possible, in which the user interacts using speech. It could be said that the main feature of these systems is their aim to behave 'intelligently' as if they were human operators in order to increase the speed, effectiveness and ease of obtaining specific services automatically. For that purpose, these systems typically include a module that implements the 'intelligence' of the human being whose behaviour they aim to replace in order to provide users with a natural and effective interaction. Among other applications, these systems have been used to provide automatic telephone services such as airplane travel information (Seneff and Polifroni 2000), train travel information (Billi et al. 1997; Torres et al. 2003; Vilar et al. 2003), weather forecasts (Zue et al. 2000; Nakano et al. 2001), fast food ordering (Seto et al. 1994; López-Cózar et al. 1997), call routing (Riccardi et al. 1997; Lee et al. 2000), and directory asistance (Kellner et al. 1997).

The use of dialogue systems has increased notably in recent years, mainly due to the important advances made by the Automatic Speech Recognition (ASR) and speech synthesis technologies. These advances have allowed the setting up of systems that provide important economic savings for companies, offering an automatic service available 24 hours a day to their customers. The initial systems were very limited with regard to the types of sentences that could be handled and the types of task performed but in the past three decades the communication allowed by this kind of system has improved notably in terms of *naturality* or similarity to human-to-human communication. However, the dialogue between human beings relies on a great diversity of knowledge that allows them to make assumptions and simplify the language used. This makes it very difficult for current dialogue systems to communicate with human beings in the same way humans carry on a dialogue with each other.

Although the functionality of these systems is still limited, some of them allow conversations that are very similar to those carried out by human beings, they support natural language phenomena such as anaphora and ellipsis, and are more or less robust against spontaneous speech phenomena as, for example, lack of fluency, false starts, turn overlapping, corrections, etc. To achieve robustness in real-world conditions and portability between tasks with little effort, there is a trend towards using simple dialogue models when setting up these systems (e.g. state transition networks or dialogue grammars) and using simple representations of the domains or tasks to be carried out. This way, it is possible to use information regarding the likely words, sentence types and user intentions. However, there are also proposals on using much more complex approaches. For example, there are models based on Artificial

Intelligence principles that emphasise the relationships between the user's sentences and his plans when interacting with a human operator or an automatic system, and the importance of applying reason to the beliefs and intentions of the dialogue partners. There are also hybrid models between both approaches that attempt to use the simple models enhanced with specific knowledge about the application domain, or that include plan inference strategies restricted to the specific application of the system.

Speech is the most natural communication means between human beings and is the most adequate communication modality if the application requires the user to have his eyes and hands occupied carrying out other tasks, as, for example, using a mouse, a keyboard, or driving a car. However, dialogue systems based exclusively on speech processing have some drawbacks that can result in less effective interactions. One is derived from the current limitations of the ASR technology (Rabiner and Juang 1993), given that even in very restricted domains and using small vocabularies, speech recognisers sometimes make mistakes. In addition to these errors, users can utter out-of-domain words or sentences, or words of the application domain not included in the system vocabulary, which typically causes speech recognition errors. Hence, to prevent these errors in the posterior analysis stages, the systems must confirm the data obtained from the users. Another problem, specially observed in telephone-based dialogue systems that provide train or airplane information, is that some users may have problems understanding and taking note of the messages provided by the systems, especially if the routes take several transfers or train connections (Claasen 2000). Finally, it has also been observed that some users may have problems understanding the possibilities of the system and the dialogue status, which causes problems of not knowing what to do or say.

1.2.1 Technological Precedents

SDSs offer diverse advantages in comparison to previous technologies developed to interact with remote, interactive applications that provide information or specific services to users. For example, one of these technologies is Dual Tone Multiple Frequency (DTMF), in which the user interacts with the system by pressing keys on the telephone that represent functions of the application (e.g. 1 = accept, 2 = deny, 3 = finish, etc.). Although this type of interaction may be appropriate for applications with a small number of functions, dialogue systems allow much more flexibility and expression power using speech, as users do not need to remember the assignation of keys to the functions of the application.

Another way to communicate with remote applications, before the advent of the current dialogue systems, is based on using speech in the form of isolated words. This way it is possible to use the words 'yes' or 'no', for instance, instead of pressing keys on the telephone. This technology has been implemented in applications with small vocabularies and a small number of nested functions. In comparison to this technology, dialogue systems offer the same advantages mentioned before, given that in these applications speech does not allow any expression facility and is a used merely as an alternative to pressing telephone keys.

Another technology developed to interact with remote applications using speech is based on the so-called *sub-languages*, which are subsets of natural language built using simple grammars (Grishman and Kittredge 1986). Following this approach, Sidner and Forlines (2002) developed a sub-language using a grammar that contains fourteen context-free rules to parse very simple sentences, mainly formed by a verb in the imperative form followed by a noun or a pronoun. This sub-language was used to interact with the Diamond Talk system developed to record and playback television programs. The system receives the user voice, processes it and generates an output both in speech form (using a speech synthesiser) and text format (on the TV screen).

Compared with this technology, dialogue systems provide a more natural interaction since users are not forced to learn specific languages: they can use their own natural language. Thus, these systems facilitate the query of databases, specially for non-expert users who can carry out queries expressed in natural language instead of having to learn database-specific languages as SQL. Also, these systems allow the users to refine and modify the queries by means of a continuous dialogue with the system, instead of having to build the queries in just one command.

1.3 Multimodal Dialogue Systems

Dialogue systems can be classified according to several criteria. One is the communication modalities (or channels) used during the interaction. Regarding this criterion, these systems can be classified into *unimodal* and *multimodal*. In a unimodal system the information is transmitted using just one communication channel; this is the case to SDSs since the interaction is carried out using only speech. These systems have several drawbacks mainly derived from the current limitations of Automatic Speech Recognition (ASR) technology. In order to solve these limitations, in the past few years there has been a notable research interest in the joint utilisation of speech and other communication modalities, leading to the so-called Multimodal Dialogue Systems (MMDSs) that are studied in Chapter 3.

Multimodality refers to the use of more than just one communication channel to transmit/receive information to/from the user (e.g. speech, gestures, gaze, facial expressions, etc.), which allows a reduction in the number of interaction errors. In fact, human-to-human communication relies on several communication modalities such as, for example, speech, gaze, body gestures, facial expressions, etc. Human beings use all these information sources (unconsciously on many occasions) to add, modify or substitute information in the speech-based communication, which allows effective recognition and understanding rates even in noisy environments. MMDSs aim to replicate the human-to-human communication which is in essence a multimodal process, given that several communications channels take part. Thus, multimodal systems are designed to support more transparent, flexible, effective and powerful interaction in a wider range of applications, to be used by non-expert users and to offer a more robust performance in changing environmental conditions. Among other applications, these systems have been developed to provide information about microwave ovens (Beskow and McGlashan 1997), available apartments (Gustafson et al. 1999, 2000; Bickmore and Cassell 2004), or boat traffic conditions (Beskow 1997), and have also been applied to the interaction with mobile robots (Lemon et al. 2001), and appointment scheduling (Vo and Wood 1996).

The so-called *multimedia* systems also use several communication channels to transmit/receive information. For example, they can transmit sounds and speech to the user through the loudspeakers, and present graphic and written information on the screen. Also, they can receive information from the user through the keyboard, or a microphone, etc. However, the difference between both types of system is that the multimodal one processes information at a higher abstraction level in order to obtain its meaning (i.e. the user intention) whereas this high-level information representation is unnecessary for multimedia systems.

In this book we consider that any dialogue system can be seen as a particular case of MMDS that uses more or less interaction modalities, depending on the type of dialogue system. Typically, the design of such a multimodal system considers the independence of dialogue management strategies, the internal representation of data provided by the user, the task to be carried out by the system, and the interaction modalities. This way a user may provide data to the system using a microphone, computer display, data glove or keyboard, for example, while the system may provide responses using any output device, as for example a loudspeaker or a display. The input provided by the user through any of the input devices is converted into an internal representation independent of the device. Thus, it makes no difference for the system where the input comes from, i.e. the user may either use a pen to click on a 'Cancel' button shown on the screen, utter a particular word associated with this action, or press the ESC key in the keyboard. This independence leads to a system architecture consisting of several modules that can be organised into three components: input interface, multimodal processing and output interface, together with the task and database modules (Figure 1.1). This figure shows the conceptual module structure we are using for explanation purposes throughout this book. The system modules are discussed with some detail in Chapter 2.

In the input interface (studied in Section 2.1) the ASR module analyses the user voice and transforms it into a sequence of words in text format. The natural language processing (NLP) module[1] analyses the sequence of words in order to obtain its meaning. The face localisation and tracking module finds the face of the user as soon as he or she enters into the vision range of a camera connected to the system, and tracks its movement in a sequence of images. The gaze tracking module processes the images provided by the camera to detect and track the movement of the user's eyes. The lip-reading recognition module detects and analyses the user's lip movements, which are very important to enhance the ASR specially when the speech signal is degraded by noise or cross-talk (i.e. other people talking near the microphone or the telephone). The gesture recognition module analyses the gestures

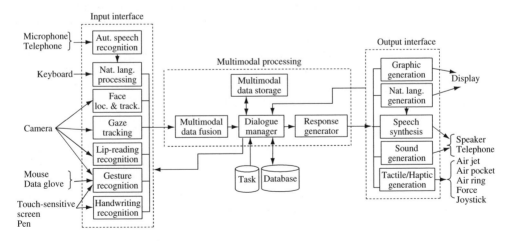

Figure 1.1 Conceptual structure of a multimodal dialogue system

---

[1] Some researchers call this module 'NLU' instead of 'NLP' as it is concerned with the Natural Language Understanding. Thus, in this book we call it either NLP or NLU.

made by the user to detect specific movements of some parts of the body (e.g. hands, arms, head, etc.) and to recognise determined communicative acts (e.g. affirm, negate, etc.). The handwriting recognition module deals with the input the user may provide by writing (using a stylus on a Personal Digital Assistant (PDA) screen, for instance). This input modality is specially important for applications that use foreign words (e.g. city names), which may be easier recognised written than spoken. In addition, this interaction modality can be used to provide additional vocabulary to the system.

In the multimodal processing component (studied in Section 2.2), the multimodal data fusion module receives the information chunks provided by the diverse input modules and combines them in order to obtain the intention of the user (operation called *fusion*). This module sends the obtained intention representation to the next module of the architecture, called the dialogue manager, that decides the next action of the system in order to carry out a particular task defined in the task module of the system. By changing the contents of this module the system can be ported to a different application domain, i.e. carry out a different task. Considering the task and the dialogue status, the dialogue manager can initiate a diversity of actions, for example, a query to the database module (either local or remote) that contains the information requested by the user, a prompt for additional data to query the database, a confirmation of data obtained from the user with little confidence, etc. The representation of the user intention is stored in the so-called multimodal data storage module, which stores all the interactions made by the user (and possibly the system) and provides the dialogue history for the dialogue manager to resolve possible ambiguities and referring expressions. Finally, the response generator module carries out the *fission* operation (as opposed to the fusion operation commented before), consisting in choosing the output to be produced in each output modality and coordinating the output across the modalities.

In the output interface (studied in Section 2.3), the graphic generation module presents information visually on the computer screen (e.g. graphics, pictures, icons, videos, etc.). It usually also includes a human-like character called the *animated agent* (studied in Section 3.2.4), that moves the lips and makes facial expressions and/or body gestures in synchronisation with the speech output. The natural language generation (NLG) module creates messages in text mode to be displayed on the screen and also text messages to be converted into the system voice through the speech synthesis module, which makes the text-to-speech conversion. The sound generation module generates specific sounds that can be useful in providing a more informative, friendly and comfortable interaction for the user, as, for example, a distinct sound if the sentence uttered cannot be understood by the system. Finally, the tactile/haptic module generates output information to stimulate the somatic senses of the user using tactile, haptic or force output devices.

It should be noted that Figure 1.1 presents a conceptual, general architecture whose modules have assigned the functions mentioned above. In some implemented systems some of these modules are subdivided into other modules, whereas in other systems, modules separated in this figure are integrated in just one module. For example, in the SmartKom system (Pfleger et al. 2003), the dialogue manager is divided into two modules: *discourse modeller* and *action planner*. The task of the former is to enrich the hypotheses generated by the data fusion module with information about the current context, validate these hypotheses, and collaborate with the fusion module in resolving referring expressions, whereas the task of the action planner is to access connected services and/or devices. However, in the Olga system (Beskow and McGlashan 1997), the dialogue manager not only decides the next action of the system but also decides and coordinates the response generation; thus, in this case the response generation module can be considered integrated into the dialogue manager.

In Figure 1.1, the arrow from the dialogue manager module to the input interface indicates that this module provides information that can be used by several modules of the interface. This information is typically concerned with the dialogue status and expectations about the input the user will likely make in his next turn, which may be very useful to enhance the performance of several recognition modules. For example, the expectations may be useful in selecting a particular language model used by the speech recognition module. Analogously, the arrow from the output interface to the dialogue manager indicates that the output of several modules in the interface can be used as feedback to the dialogue manager. For example, this information may be useful to resolve user references to objects previously shown on the system screen (studied in Section 5.4.5).

It can also be noted that considering the system architecture shown in Figure 1.1, an SDS can be considered as a particular case of an MMDS in which only the modules concerned with speech processing (ASR, NLP, NLG and speech synthesis), in addition to the dialogue manager, are used. In this case, the input is unimodal (only speech), the multimodal data fusion only considers data obtained from the uttered sentences, and the generation of responses is also unimodal (only speech and maybe other sounds).

1.4 Multilingual Dialogue Systems

Another criterion to classify dialogue systems is in terms of the number of languages supported. Considering this criterion, they can be classified into *monolingual* and *multilingual*. The latter solves a limitation of the former, namely that the interaction must be carried out using a specific language, which makes them unusable for users who do not speak that language. Multilingual Dialogue Systems (MLDSs), studied in Chapter 4, present the advantage of allowing the interaction using several languages (e.g. English, French, German, etc.). Among other applications, these systems have been applied to provide information about geographical areas, such as points of interest, distance between places, etc. (Glass et al. 1995), and weather information (Zue et al. 2000). In particular, these systems are very useful for some types of user, e.g. tourists who cannot speak the local language of the country they are visiting. Generally these systems are designed to be as language-independent as possible by storing language-dependent information in external configuration files. This development is based on the fact that it is possible to obtain the same semantic representation from the analysis of a given sentence expressed in different languages, in which only the language-dependent words change. This method of analysing sentences is similar to the *interlingua* approach to Machine Translation (MT) (Roe et al. 1991; Waibel et al. 1991; Hutchins and Somers 1992; Wahlster 1993, 1994). Given that the semantic representations do not change, the dialogue management strategies can be language-independent. Also, instead of using just one set of patterns to generate the sentences, these systems generally use several sets (one per language) which allow to generate the same kind of spoken sentences in different languages using a multilingual NLG module (for generating the sentences in text mode) and a multilingual speech synthesiser (to carry out the text-to-speech conversion).

1.5 Dialogue Systems Referenced in This Book

A large amount of dialogue systems (either spoken, multilingual or multimodal) for a great number of application domains can be found in the literature. Table 1.1 shows some of them, which have been mostly referenced throughout this book for explanation purposes.

Table 1.1 Dialogue systems mostly referenced in this book

System/Year/Reference	Application domain	Lab	Input	Output	Languages	Remarks
AdApt Gustafson et al. 2000, 2002; Bell et al. 2000; Skantze 2002	Real estate	KTH and Telia Research (Sweden)	Speech, mouse, gestures	Speech, graphics, animated agent	Swedish	
August 1999 Gustafson et al. 1999	Info. about city facilities, KTH and life and works of A. Strindberg	KTH (Sweden)	Speech	Speech, text, animated agent	Swedish	Push-and-talk for ASR
Jeanie 1996 Vo and Wood 1996	Appointment scheduling	Carnegie Mellon Univ. (USA)	Speech, gestures, handwriting	Speech	English	
Jupiter 2000 Zue et al. 2000	Weather information	Lab for Computer Science MIT (USA)	Speech	Speech	English, German, Japanese, Mandarin Chinese, Spanish	
KIT system 2002 Xu et al. 2002	Sightseeing, hotel search, PC assembly	KIT (Japan)	Speech	Speech	Japanese, Chinese	
Kyoto voice portal 2005 Omukai and Araki 2004	Weather, bus, restaurant, sightseeing	KIT (Japan)	Speech	Speech, graphics	Japanese, English	

Matis 2000 Sturm et al. 2000	Train timetable travel information	Dept. Language and Speech, Nijmegen (The Netherlands)	Speech, graphics	Speech, graphics	Dutch	
MUST 2002 Almeida et al. 2002	Tourist information	Portugal Telecom Inovação, Telenor R&D, France Télécom R&D, Univ. Nijmegen, Max Planck Inst. Psycholinguist.	Speech, pen pointing	Speech, text, graphics	Norwegian French Portuguese English	Mobile
Olga 1997	Microwave ovens	KTH and Institute of Computer Science (Sweden)	Speech, direct manipulation via widgets	Speech, animated agent, graphics	Swedish	
Philips 1995 Aust et al. 1995	Train timetable travel information	Philips Speech Processing (Germany)	Speech	Speech	German	
QuickSet 2000 Oviatt 2000	Interaction with distributed simulations	Oregon Graduate Institute (USA)	Speech, pen gestures	Graphics	English	Mobile
Rea 2000 Bickmore and Cassell 2004	Real state	MIT Media Lab (USA)	Speech, hand gestures	Speech, animated agent, graphics	English	

(continued overleaf)

Table 1.1 (*continued*)

System/Year/Reference	Application domain	Lab	Input	Output	Languages	Remarks
Schisma 1996 Van de Burgt et al. 1996	Theatre performance	KPN & Univ. of Twente (Netherlands)	Speech	Speech	Dutch	
SmartKom 2000	Operation of machines at home, access to public services,	DFKI (Germany)	Speech, hand gestures, facial expressions	Speech, animated agent, graphics	English German	Mobile
TeleTuras 2004	Tourist information	University of Ulster (Northern Ireland)	Speech, text, haptic deixis	Graphics, text, audio	English	Mobile
Voyager 1995 Glass et al. 1995	regional information	Lab for Computer Science MIT (USA)	Speech	Speech, graphics	English, Japanese, Italian	
Waxholm 1993 Carlson and Granström 1993	Boat traffic in Stockholm archipelago	KTH (Sweden)	Speech	Speech, graphics	Swedish	
Witas 2001 Lemon et al. 2001	Dialogue with autonomous robots	CSLI, Stanford Univ. (USA)	Speech	Speech, graphics	English	Complex dialogue management

1.6 Area Organisation and Research Directions

The research on dialogue systems is a very interdisciplinary field that covers a wide range of research areas, such as speech perception, signal processing, phonetics, phonology, NLP, discourse processing, dialogue management, natural language generation (NLG), speech production, prosody, user modelling, knowledge representation, human-computer interaction, evaluation, standardisation, etc. Consequently, these systems are addressed, with more or less specialisation, by the research associations concerned with these topics, as for example, ACL[2] (Association for Computational Linguistics), ACM[3] (Association for Computing Machinery), AAAI[4] (American Association for Artificial Intelligence), ASJ[5] (Acoustical Society of Japan), ELRA (European Language Resources Association), ISCA[6] (International Speech Communication Association), IEEE[7] (Institute of Electrical and Electronic Engineers), JSAI[8] (Japanese Society for Artificial Intelligence), SEPLN[9] (Spanish Society for Natural Language Processing), etc. A number of conferences and workshops are organised regularly by these associations including topics concerned directly with dialogue systems, or with some of the technologies employed to build them, as for example:

- AAAI conferences
- ACL meetings
- CHI (Computer-Human Interaction) conferences
- ISCA Eurospeech/Interspeech conferences
- ICASSP (International Conference on Acoustics, Speech, and Signal Processing)
- IEEE ASRU (Automatic Speech Recognition and Understanding) workshops
- ICSLP (International Conferences on Spoken Language Processing)
- SEPLN (Spanish Society for Natural Language Processing) conferences
- SIGdial[10] Workshops (sponsored by ACL and ISCA)
- SIGMEDIA Workshops
- Workshops on knowledge and reasoning in practical dialogue systems, held during the IJCAI (International Joint Conference on Artificial Intelligence).

The above-mentioned research institutions, among others, stimulate considerably the advances in the field, promoting the transfer of prototype systems and research results to industrial applications. They also promote the publication of books and papers in a variety of journals, such as *Speech Communication*,[11] *Computer Speech and Language*,[12] *Signal Processing*,[13] etc. Several companies are also very interested in this technology, researching,

---

[2] http://www.aclweb.org
[3] http://www.acm.org
[4] http://www.aaai.org
[5] http://www.asj.gr.jp/index-en.html
[6] http://www.isca-speech.org
[7] http://www.ieee.org/portal/site
[8] http://www.ai-gakkai.or.jp/jsai/english.html
[9] http://www.sepln.org
[10] http://www.sigdial.org
[11] http://ees.elsevier.com/specom
[12] http://ees.elsevier.com/csl
[13] http://www.ijsp.org/topics.htm

offering and commercialising products to ease the design, development and evaluation, such as IBM,[14] Microsoft,[15] Nuance,[16] Philips,[17] ScanSoft,[18] etc.

There is a variety of current research directions in the field of dialogue systems, addressed in the mentioned journals, conferences and workshops. For example, one line of research is concerned with the development of tools to facilitate the implementation, not only for systems to be set up in desk-top computers but also to be set up in mobile devices. The extension of wireless communication to the general public has caused a notable increase in services and applications to provide information at any time and anywhere. Among other services, dialogue systems whose input is carried out using speech or pen-based modalities are particularly useful for tasks carried out on the move, such as personal communication or navigation in tourist applications. However, the setting up of dialogue systems for mobile devices implies problems not addressed for systems set up in desk-top devices. These problems represent another line of research that has been addressed by several projects. Among others, the TeleMorph project (Solon et al. 2004) has developed a framework for setting up MMDSs considering the fluctuations of the available bandwidth. Considering this factor, the multimodal output is decided dynamically using specific modalities.

Another research line is concerned with the design of more robust systems, which can be successfully used in the real world, considering both the environmental conditions and the users. Research work is being carried out to increase the robustness of all the technologies employed to set up these systems (e.g. ASR, NLP, dialogue management, speech synthesis, etc.). For example, in terms of dialogue management, work has been carried out to set up adaptive techniques that learn user preferences and adapt the dialogue strategies to the current user, in an attempt to improve the task completion and the user satisfaction. For example, Singh et al. (2002) used reinforcement learning techniques to determine the system's level of initiative and the amount of confirmation for the user utterances. Other researchers have proposed adapting the dialogue strategies over the course of the conversation based on user responses or other dialogue characteristics. For example, Litman and Pan (2000) presented an adaptive technique based on rules to decide whether the user interaction was becoming problematic. In such a conflict situation, their method changes the interaction strategy from mixed-initiative to system-directed strategy (Section 5.3.1) and the confirmation strategy from implicit to explicit (Section 5.3.2). Taking a different approach, Maloor and Chai (2000) presented a help-desk application that first classifies the user as a novice, moderate or expert based on responses to prompts, and then adjusts the complexity of system utterances, the vocabulary, and the complexity of the path taken to achieve the goals during the dialogue.

The development of emotional dialogue strategies represents another line of research, which relies on the fact that emotions play a very important role in the rational decision-making, perception and human-to-human interaction (Section 3.2.5). This fact has created the interest in implementing the ability to recognise and reproduce emotions in computers. From a general point of view, emotional dialogue strategies take into account that human beings usually exchange their intentions using both verbal and non-verbal information (Cassell and

---

[14] http://www-306.ibm.com/software/voice
[15] http://www.microsoft/speech
[16] http://www.nuance.com
[17] http://www.speechrecognition.philips.com
[18] http://www.scansoft.com

Thorisson 1999). In fact, although we as humans sometimes do not express our intentions explicitly using words, our conversation partners can recognise our intentions through our face expression and/or some features of our voice. Thus, in SDSs, several researchers have used voice prosody features to detect negative or non-negative emotions. For example, Lee et al. (2001) used ten features representing the voice fundamental frequency and energy to enhance the quality of service in call centre applications. Similarly, Ang et al. (2002) used durations, speaking rate, pause, pitch and other prosodic features to detect frustration and annoyance in natural human-computer dialogue. Forbes-Riley and Litman (2004) used acoustic-prosodic and other types of linguist features, extracted from the current turn and a context of previous user turns, to predict the user emotion and automatically adapt the system accordingly. In MMDSs, a variety of methods have been proposed to detect the user emotional state and adapt the system behaviour, from methods based on artificial neural networks (ANNs), Hidden Markov Models (HMMs), FACS or MPEG-4[19] to capture the user's expression in pictures or video sequences, to methods to capture the user emotional state based on biosensors connected to the user body.

As discussed in Section 3.2.5.2, the development of *social* dialogue strategies is another research direction. It relies on the fact that in human-to-human dialogues people not only speak about topics concerned with the task at hand, but also about topics not concerned with the task (e.g. weather conditions, current news, etc.), especially at the beginning of the conversation. Several researchers indicate that this feature of human dialogue is a result of social behaviour. Among other questions, this behaviour can be very useful to know the objectives and plans of the interlocutor, as well as his or her way of expression and attitudes towards the task. For example, salesmen typically use this *social* dialogue to get acquainted with the clients, and to control when and how to introduce the information concerned with the task. Several authors indicate that non-verbal information plays a very important role in this dialogue type. For example, Whittaker and O'Conaill (1997) showed that non-verbal information provided in a video-mediated communication did not influence the task-oriented dialogue interactions, while it had a clear impact on the dialogue interactions concerned with getting acquainted or negotiation. This result suggests that setting up dialogue systems that behave more like human beings requires including abilities that are typical of social behaviour, such as the ability to decide whether to speak about topics not concerned with the task at hand (e.g. the weather), or about topics related to the task (e.g. money to pay for an apartment rent).

1.7 Overview of the Book

In addition to this introductory chapter, the book is divided into six chapters. Chapter 2 describes the diverse modules of the input interface, multimodal processing and output interface of the conceptual dialogue system shown in Figure 1.1. It initially addresses the ASR module, including a little bit of history, classifications of ASR systems, main problems, types of error, the stochastic approach to ASR, and the acoustic and language modelling. Then the chapter focuses on the NLP module, discussing basic analysis techniques, typical problems and suitability of some grammars for dialogue systems. Third, the chapter addresses the modules concerned with face localisation and tracking, and gaze tracking, describing methods

---

[19] FACS (Facial Action Coding System) and MPEG-4 are discussed in Section 3.2.4.1 for face animation.

to carry out these tasks. The next section focuses on the lip-reading recognition module and discusses recognition techniques. The chapter then addresses the gesture recognition module, discussing classifications of gestures and recognition methods, and focuses on the handwriting recognition module, discussing several recognition approaches. After studying the input interface, the chapter deals with the multimodal processing component, addressing the modules concerned with the multimodal data fusion and storage, as well as the modules for dialogue management, task representation, database and response generation. The chapter ends by describing the output interface. It initially addresses the graphic generation module, giving a brief introduction to animated agents and other graphic information. Then it focuses on the NLG module, discussing the traditional phases and methods developed to generate text sentences. The chapter then briefly addresses the speech synthesis module, discussing a little bit of history and the main speech synthesis techniques. Finally, briefly the sound generation and tactic/haptic modules are presented, with the use of *auditory icons* as well as available devices for tactic/haptic generation discussed.

Chapter 3 deals with Multimodal Dialogue Systems (MMDSs) and is divided into two parts. The first presents a discussion of the benefits of MMDSs in terms of system input, processing and output, whereas the second focuses on development concerns. The latter addresses two typical development techniques (WOz and system-in-the-loop) and focuses on the problem of multimodal data fusion, addressing typical kinds of such a process. Next it addresses architecture concerns and discusses two architectures commonly used in the setting up of MMDSs. The fourth section of this second part is concerned with the animated agents. The section initially addresses several approaches to facial animation and then discusses face and body animation standards and languages. The chapter concludes by discussing some recent trends concerned with the design of MMDSs, focusing on emotional dialogue, personality and social behaviour.

Chapter 4 deals with Multilingual Dialogue Systems (MLDSs). The first section initially addresses the implications of multilinguality in the architecture of dialogue systems, focusing on speech recognition and synthesis, language understanding and generation, level of interlingua, and dialogue content dependency. Then, it addresses three approaches to the implementation (interlingua, semantic frame conversion and centred on dialogue control). The second section then describes some research systems based on interlingua, in particular focusing on how to set up language independency in semantic representations, how to deal with the differences in each language, and how to generate multilingual sentences. The third section explains the advantages and disadvantages of the architecture based on the Web application framework, and how to construct a multilingual dialogue system following the internationalisation methodology, which is widely used in GUI-based Web applications.

Chapter 5 focuses on dialogue annotation, modelling and management. It initially addresses the annotation of spoken dialogue corpora at the morphosyntactic, syntactic, prosodic and pragmatic levels. Then it discusses the annotation of multimodal dialogue corpora, comments on some efforts made to promote the development of guidelines and the corpora sharing, and gives a brief introduction to the annotation of speech, gestures and other interaction modalities. In the second section it addresses the dialogue modelling, focusing on the dialogue models based on state-transition networks and plans. Third, it addresses the dialogue management techniques, focusing on the interaction and confirmation strategies. In the fourth section it addresses the implications of multimodality in dialogue management in terms of interaction complexity, confirmation strategies, social and emotional dialogue, contextual information, user references, and response generation. The next section addresses the implications of multilinguality in dialogue management, discussing reference resolution, ambiguity of speech acts, and the differences in the interactive behaviour of the systems. The final section of the chapter deals with the implications of task independency in dialogue management, focusing on the dialogue task classification and the task modification in each task to achieve task independency.

Chapter 6 addresses development tools. It is divided into two sections. The first discusses tools to develop spoken and MLDSs, while the second addresses tools to develop MMDSs. The first section focuses initially on tools to develop system modules, addressing the ASR (HTK), language modelling (CMU – Cambridge Statistical Language Modelling Toolkit) and speech synthesis (Festival, Flite, MBROLA and Galatea talk). Then it presents Web-oriented standards and tools to develop SDSs, including VoiceXML, CCXML, OpenVXI, publicVoiceXML and IBM WebSphere Voice Toolkit. Then it discusses some available Web portals for setting up SDSs (Voxeo Community, BeVocal Café and VoiceGenie Developer Workshop). The second section of the chapter presents tools for setting up MMDSs. It first addresses the Web-oriented multimodal dialogue, studying XHTML + Voice and SALT. Then it briefly presents some tools for face and body animation (Xface Toolkit, BEAT and Microsoft Agent). The next section previews some toolkits to set up complete MMDSs (CSLU Toolkit and IBM Multimodal Tools). The chapter concludes by discussing tools for annotating multimodal corpora (Anvil and Tycoon).

Finally, Chapter 7 focuses on the assessment of dialogue systems. The chapter gives an initial introduction to the evaluation techniques for dialogue systems, classifying them into four categories: subsystem-level, end-to-end, dialogue processing and system-to-system. The second section of the chapter addresses the evaluation of spoken and MLDSs considering these four approaches. It initially focuses on the subsystem-level evaluation, discussing the assessment of the ASR, NLU, NLG and speech synthesis modules. The section also addresses the end-to-end and the dialogue processing evaluation, describing the widely used PARADISE framework, and commenting on an analytical approach for evaluating dialogue strategies. Next, the section describes the system-to-system evaluation, discussing the assessment with limited resources, linguistic noise and a simulation technique. The third section of the chapter addresses the evaluation of MMDSs. It initially focuses on the system-level evaluation, discussing the experimental, predictive and expert-based approaches, and commenting briefly on the PROMISE evaluation framework. Finally, it presents the subsystem-level evaluation, discussing initially the assessment of the modules concerned with face localisation and gaze tracking, gesture recognition and handwriting recognition. Then, it addresses the multimodal data fusion, focusing on the fusion of speech and gestures, speech and visual information, and the three modalities together. Next, the chapter addresses the evaluation of the animated agents, studying the effects they produce on the system's effectiveness and naturalness, intelligibility, and on the user's comprehension and recall. Then is presents the effects on the user of the agent's facial expressions and its similarity with the human behaviour. To conclude, the chapter discusses as methodological framework to guide the evaluation of animated agents.

1.8 Further Reading

- Part I of Huang et al. (2001).
- Part I of McTear (2004).

2

Technologies Employed to Set Up Dialogue Systems

2.1 Input Interface

The first computers developed for general public use were based on keyboards and CRT screens. Traditional keyboards allow text input and several basic actions (e.g. Enter/Escape), navigation (e.g. Tab, Home, End) while modern keyboards allow more sophisticated operations (e.g. MIDI, start of applications, etc.). Despite its ease and widespread use, a keyboard is not the best device for interacting with any kind of application, since some users cannot use it due to physical limitations, or because they do not have the skill to do so. Moreover, using a keyboard is particularly difficult in some situations, for example, when driving a vehicle.

Particularly with graphic user interfaces (GUIs), the mouse is typically used as the most common input device for positioning and selecting tasks. This device can be of different shapes and be equipped with one, two or three buttons, working either mechanically or optically. In addition to keyboards and mice, current computers employ other devices such as microphones, loudspeakers, video cameras, modems, scanners, joysticks, etc. In general, it can be said that the best device for a particular application depends on the task to be performed when the user interacts with the application. In the case of MMDSs, input devices can be classified into several categories: pointing (e.g. mouse, touch-sensitive screen, data glove, etc.), sound (e.g. microphone, loudspeakers, etc.), image (e.g. displays, cameras, eyetrackers, etc.), and other tactile/haptic devices (e.g. pneumatic, vibrotactile, electrotactile, etc.). Mice are generally used for positioning purposes. Cameras are employed to obtain images of the user that are generally analysed for lip-reading, gesture recognition, facial expression recognition, face localisation and face tracking. Microphones are used to obtain the user voice to be analysed by a speech recogniser, while touch-sensitive screens are used to allow the user can point to objects in selection tasks.

After this brief discussion of input devices, this section now focuses on the input modules of Figure 1.1. Given that each input module represents a very wide research area, we do not have enough space in this chapter to examine all the modules in detail, but rather present

Spoken, Multilingual and Multimodal Dialogue Systems: Development and Assessment   Ramón López-Cózar Delgado
and Masahiro Araki   © 2005 John Wiley & Sons, Ltd

a general overview that serves as an introduction to further studies. Section 2.5 provides references for further reading about the input interface modules.

2.1.1 Automatic Speech Recognition

The human voice is originated from the vibration of the vocal cords. This vibration moves as sound waves in all directions in the air, from the speaker's larynx to the listener's ear. The ear receives the pressure of the sound waves and transforms it into a signal that is sent to the brain, where it is interpreted as a sound. The acoustic features of this signal allow the listener to differentiate one sound from another. Therefore, the goal of an ASR system is to determine the sound units considered in the speech signal, either phonemes or larger units (e.g. syllables), and combine them to determine the words uttered by the speaker. Although apparently simple for human beings, it is a very complex and difficult task for the current ASR technology since the human ability to recognise speech relies on the use of a variety of knowledge, such as lexical, syntactic, semantic, pragmatic features, and about the world in which we live.

2.1.1.1 A Brief History

The beginning of ASR can be traced back to 1870 when Alexander Graham Bell tried to build a machine able to represent words visually. Bell was not successful in his proposal but his work was the origin of the telephone. In 1952, AT&T Labs developed a system able to recognise speech (concretely the ten digits) using an analogic computer. The system had to be trained with the same voice that it was to recognise, obtaining a recognition accuracy of 98%. Later, a system was developed to recognise consonants and vocals. Conceptually, a speech recogniser is comprised of two modules, namely the *feature extractor* and the *classifier* (Rabiner et al. 1996). The former receives the voice signal, divides it into a sequence of fragments, and applies to them a signal processing technique to obtain a representation of the most significant features. Considering these features, a sequence of vectors is built that is the input for the classifier. This module builds a probabilistic model based, e.g. on ANNs or HMMs, carries out a search to find the most likely segment sequence, and provides a sequence of recognised words. In order to recognise the user voice, ASR systems must be appropriately trained in advance.

2.1.1.2 Classifications of ASR Systems

These systems can be classified according to several criteria. For example, in terms of allowed speakers they can classified into *single speaker*, *speaker independent* or *speaker adapted* systems. In the former case, the system can only recognise words uttered by the same speaker who trained it. This type of system obtains high recognition rates and is the one to use if the speaker has specific pronunciation problems. Speaker independent systems can recognise words uttered by other users in addition to the training ones. In general, these systems obtain lower recognition rates than the single-speaker systems, and are the ones typically used for telephone-based applications. Finally, in the case of speaker-adapted systems several speakers undertake the initial training, and in order to carry out the recognition the systems are adapted to a particular user.

In terms of allowed user speaking style, ASR systems can be classified into *isolated words*, *connected words*, and *continuous speech* systems. The former require the speakers to make clearly defined pauses between words, which facilitates the detection of the start and end of words. Connected word recognition systems require the speakers to make short pauses between the words, while continuous speech recognition systems do not require the speakers to make pauses to delimit words, allowing them to speak normally. In the latter there is no distinction regarding the end of words (e.g. consider the sentence 'I scream for ice-cream'). Also according to speaking style, there are systems that can only handle a formal speaking style, typically called *read speech* systems, while other systems are trained to handle *spontaneous speech*, allowing to speakers talk as naturally as if they were talking to another person. These systems are much more complex since they must face typical problems of this kind of speech, such as hesitation, repetition, ungrammatical sentences, etc. These are the systems typically used in current SDSs, although previous systems have used isolated-word recognition, a for example, the Railtel system to provide train information (Billi et al. 1997).

There is another type of ASR systems, called *keyword spotting* systems, which do not attempt to obtain the accurate sequence of words uttered by the speaker, but only determined words, the others being considered irrelevant (Heracleous and Shimizu 2003). The first systems of this kind appeared in the 1970s and were based on the Dynamic Programming technique, whereas at the end of the 1980s and the beginning of the 1990s, initial systems based on HMMs were presented.

The task of ASR is to analyse the user voice and transform it into a sequence of recognised words. A variant of this task is the so-called *N-best* recognition (Stolcke et al. 1997). In this case the recogniser does not provide a sequence of recognised words but N (one optimal and N-1 suboptimal). This technique is useful because sometimes the correct word sequence is not the most likely one but is in the N-best sequence. Implementing this technique requires an increase in computation and memory requirements.

The *vocabulary* is the set of words that can be recognised. The larger the vocabulary, the greater the difficulty in recognising words, and also the larger the memory and computation requirements. In terms of this parameter, ASR systems are typically classified into three categories: *small* vocabulary systems can recognise up to 99 different words; *medium* vocabulary systems can recognise up to 999 words; and *large* vocabulary system can recognise more than 1,000 different words. There are other possible classifications, for example, some authors consider small for less than 20 words, and large for more than 20,000 words.

An important issue to consider when using ASR systems is the activation mode. The two possible methods are *automatic* and *manual* activation. The former is typically the one used for telephone-based SDSs. This kind of activation can be carried out either using 'rigid' turns, in which case the recogniser is not ready to work while the system output is being performed. The activation can also be carried out using the *barge-in* technique, in which case the recogniser starts working as soon as the user starts to speak. A variant of this technique is the so-called *hotword recognition*, where only the sentence that matches an active grammar can interrupt the prompt being played by the system. In the case of manual activation, the speaker must indicate somehow that he is going to start to talk, for example, he can either press a button (called 'Push & Talk' technique) or start the sentence with a determined word (e.g. 'computer', 'mother', etc.). This activation mode is more reliable and robust since it helps avoid the so-called *false alarms*, discussed below.

2.1.1.3 ASR Problems

ASR is a very complex task for the current technology due to a variety of reasons. The first reason is the *acoustic variability* since each person pronounces sounds differently when speaking. Even the same person cannot pronounce the same word exactly twice due to variations in time, energy, etc. A second reason is the *acoustic confusion* caused because some words sound very similar, which makes it very difficult to distinguish them (e.g. 'wall' vs. 'war', 'word' vs. 'world', etc.). Another problem is the *linguistic variability*, i.e. the distinct dialects of a language make it very hard for these systems to recognise the words uttered by determined speakers. The *coarticulation* problem is also an important factor which makes it difficult for the system to recognise the words because the features of the uttered sounds may be changed by the neighbour sounds. Another problem is the fact that the speakers may utter words not included in the system vocabulary, which will provoke recognition errors. These words, called *out-of-vocabulary* words, are typically deleted or substituted by other words included in the vocabulary, or can even provoke the insertion of non-uttered words.

Environmental conditions also impose difficulties for ASR systems. Recognition rates are typically high if the speakers talk in non-noisy lab conditions, however, when they talk in real-world conditions, the speech signal can be degraded by different factors such as music, traffic, air conditioning, *cross-talk* (i.e. other people talking near the microphone,) etc., which make it much harder to recognise the uttered words correctly. Also of note are the distortions caused by the communication channel, features of the microphones, bandwidth limitations of the telephone line, as well as other user-produced phenomena such as breaths, coughs, etc. which again make the ASR process harder.

2.1.1.4 ASR Errors

All these problems typically provoke errors in the recognition process, which are classified into five types: *insertions*, *deletions*, *substitutions*, *rejections* and *false alarms*. Insertions occur when the recognised sentence contains more words than the uttered sentence. Deletions occur when some words uttered by the speaker are missed in the recognised sentence. Substitutions occur when some words in the uttered sentence are substituted by other words in the recognised sentence. Rejections occur when the user talks but the system does not provide any output, while false alarms occur in the contrary case, i.e. when the speaker does not talk but the recogniser outputs words.

Rejection and substitution errors can occur in a variety of circumstances. For example, when the speaker utters words not included in the system vocabulary, utters sentences that are not accepted by the recognition grammar, starts talking before the system is ready to recognise (if the *barge-in* feature is not used), utters words that sound very similar, makes long pauses when uttering words, hesitates when talking (false starts, 'um', 'eh', …), or if the microphone is not set up correctly, etc. False alarms may also occur in several occasions, for example, when there are laughs, coughs, other people talking near the microphone, etc.

ASR systems are useful if high recognition rates are provided (typically at least 90%). Thus, taking into account the limitations mentioned, there are tasks that are suitable for ASR and tasks that are less appropriate for this technology. For example, ASR systems are suitable (in principle) if the keyboard and mouse are not available due to the features of the task. Also, they are a good solution if other tasks require the speaker to use his hands (e.g. driving a car), or if the speakers are not accustomed to using a keyboard or a mouse, or they have some disability (e.g. cannot move their hands). On the contrary, in general, ASR systems are not the best solution if the application requires that the speakers talk where other people are near the microphone/telephone, or if the task must be carried out in noisy environments.

In order to deal with recognition errors, SDSs systems typically use the so-called *confidence scores* provided for the ASR modules, which provide an estimation of the correct recognition of the words (Liu 2003; Cerisara 2003). Using these scores, each recognised word is assigned a value, typically a real number in the range (0–1), which represents how confident the recogniser is about the correct recognition of that word. There are several ways to compute a confidence score for a recognised word w, $CS(w)$. A simple way is based on the N-best list provided by the recogniser, for example, using the expression:

$$CS(w)_{N\text{-}best} = \frac{\sum\limits_{i=1:w \in h_i}^{N} P(h_i)}{\sum\limits_{i=1}^{N} P(h_i)} \qquad (2.1)$$

where $P(h_i)$ is the probability of the hypothesis i in the N-best list, the numerator is the sum of the probabilities of the hypothesis containing w, and the denominator is the sum of the probabilities of the N hypotheses. The idea behind this method is that a word w present in many recognition hypotheses must have a high confidence score.

2.1.1.5 The Stochastic Approach to ASR

Several approaches to the problem of ASR can be found in the literature, such as expert systems, pattern matching or stochastic (probabilistic) (Rabiner et al. 1996). The stochastic approach is currently the most used. In this approach, the ASR problem can be stated as follows: 'Given a sequence of acoustic data A, find the sequence of words W uttered'. According to this approach, this sequence is the one that verifies:

$$W = \max_{W} P(W|A) \qquad (2.2)$$

i.e. we must obtain the sequence of words with the highest probability given the acoustic data. Since it is not easy to compute $P(W|A)$, the Bayes rule is used to ease the computation, thus obtaining the expression:

$$P(W|A) = \frac{P(A|W)P(W)}{P(A)} \qquad (2.3)$$

In this expression, $P(A|W)$ is called the *acoustic model*, which represents the probability of obtaining the acoustic data A when the word sequence W is uttered. $P(W)$ is called the *language model*, which represents the probability of uttering the word sequence. Given that in this expression the probability of the acoustic data $P(A)$ is not necessary (since it is independent of the word sequence), the previous expression can be rewritten as follows:

$$W = \max_{W} P(A|W)P(W) \qquad (2.4)$$

which is the ASR fundamental equation in the stochastic approach. As mentioned above, the ASR requires two stages: training and recognition. The training is used to make the ASR system learn the acoustic and language models using voice samples and text, which reduces the recognition errors (Macherey and Ney 2003; Sasou et al. 2003). The recognition stage is the process of obtaining the output from the input, i.e. the input acoustic data (speaker sentence) is transformed into the word sequence by applying the fundamental equation of ASR (Rabiner and Juang 1993).

2.1.1.6 Acoustic Modelling

The acoustic modelling deals with the units to use in the recognition process. Choosing one or another strongly influences the performance of the recogniser. Two possibilities are word models or subword models. The former are appropriate for isolated word recognition and when the vocabulary is very small. In this case, the smallest unit that can be recognised is a word. On the contrary, subword models are appropriate for continuous-speech recognition or when the vocabulary is very large. In this case, smaller units than words are the basic recognition units, such as phonemes or syllables. Diphonemes and triphonemes (discussed in Section 2.3.3) can also be used, providing better results since they include contextual information that models coarticulation effects.

If the recognition system is based on HMMs, an HMM can be trained for each recognition unit (word or subword unit). A model is built for each acoustic unit, and a sentence is modelled as the concatenation of the models corresponding to these units. Nowadays, HMMs are mostly used to model the acoustic units, taking into account the probabilistic theory. They are typically represented using states and transitions between the states, considering that each state emits symbols or vectors with a given probability. They are said to be *hidden* because the transitions between states are not directly related to the emitted symbols.

There are three main problems that can be addressed by HMMs. On the one hand, *evaluation*, which is concerned with finding the probability of a sequence of observations given an HMM. The Forward Algorithm was designed to address this problem. Second, the *decodification* problem, concerned with finding the most likely sequence of hidden states given determined observations. The Viterbi algorithm was developed to address this problem. Finally, the *learning* problem, related to the generation of an HMM from a sequence of observations. The Forward-Backward algorithm was developed to address this.

2.1.1.7 Language Modelling

We have previously discussed the acoustic modelling, which determines the best sequence of acoustic units for a given sentence. The language modelling determines the words that can be followed by other words in the recognised sentence (Mori et al. 2003; Zitouni et al. 2003). Grammars are used to decide the word sequences. There are two basic types of grammar: *stochastic* and *rule-based*. The former are typically used for dictation applications. Stochastic grammars indicate that a word can be followed by any other word in the vocabulary with a given probability. On the contrary, rule-based grammars (also called *finite-state grammars*) indicate which words can be followed by other words using grammatical rules. Stochastic grammars calculate the language model, $P(W)$, by the following approximation:

$$P(W) = P(w_1, w_2, \ldots, w_q) = P(w_1)P(w_2|w_1)P(w_3|w_1 w_2) \ldots P(w_q|w_1 \ldots w_{q-1}) \quad (2.5)$$

However, since calculating these probabilities is difficult, it is usual to consider histories of n words, called n-grammars, where the probability of a word is calculated considering the n preceding words and assuming statistical independence from the oldest part (i.e. the preceding words) thus leading to:

$$P(w_i|w_1, w_2, \ldots, w_{i-1}) = P(w_i|w_{i-n+1} \ldots w_{i-1}) \quad (2.6)$$

Generally, n = 2 (in which case the grammar is called *bigram*) or n = 3 (in which case the grammar is called *trigram*). In the case of a bigram, calculating the probability of a word only requires considering the previous word. Thus, using a bigram, the probability of a sentence is expressed as follows:

$$P(W) = P(w_1, w_2, \ldots, w_q) = P(w_1)P(w_2|w_1)P(w_3|w_2) \ldots P(w_q|w_{q-1}) \quad (2.7)$$

The other kind of grammars are the rule-based grammars. In this case, the rules of formal grammars indicate the words that can follow a given word. These rules allow fast recognition if they are carefully designed to consider all the possible sentences the users may utter. For example, let us take a look at this sample grammar in Java Speech Grammar Format (JSGF):

```
#JSGF v1.0
// Define de grammar name
grammar SimpleCommands;
// Define de rules
public <Command> = [<Polite>] <Action> <Object> (and <Object>)*;
<Action> = open | close | delete;
<Object> = the window | the file;
<Polite> = please;
```

This grammar indicates that any sentence has a non-terminal optional symbol called 'polite' followed by other non-terminal symbols ('action' and 'object') and then the terminal symbol 'and' followed by the non-terminal symbol 'object'. These last two symbols may be repeated zero or more times. The 'action' non-terminal symbol can be transformed into the terminal symbols: 'open', 'close' or 'delete'. The 'object' non-terminal symbol can be transformed into the terminal symbols: 'the window' or 'the file'; and The 'Polite' non-terminal symbol can be transformed into the terminal symbol 'please'. Some sample sentences accepted by this grammar are 'Please open the window', 'Please open the file and the window', etc.

2.1.2 Natural Language Processing

NLP can be defined as a set of techniques to automate the human ability to communicate using natural languages. A natural language can be defined as the language used by human beings to transmit knowledge, emotions, etc. The person who wish to transmit a message codifies it using chains of sounds or written symbols, and the person who receives it decodes the chains to obtain the original message. The human ability to communicate using natural language is extremely complex, and relies on a process that builds complex structures from other initial, simple structures (written symbols are used to build words, words are used to build sentences, and the sentences are used to build the messages to transmit). In this book we will understand NLP as the second aspect of this transmission process, i.e. obtaining the coded message using written symbols or spoken words. Also, we will consider in this book that the transmission of the message is concerned with NLG.

NLP is applied in several fields of Computer Science as, for example, text processing and stylistic correction, intelligent writing assistance (detection of writing errors), database interfaces (query of databases using natural languages instead of computer-oriented languages), information extraction (to obtain information e.g. from the Internet entering queries in a browser), etc. Regarding dialogue systems, which are the focus of this book, NLP is used to obtain the semantic content of the sequence of words provided by the ASR module.

NLP can be carried out using several approaches (Dale et al. 2000). It is typical to divide the processing into several analysis types that can be performed either in an ascending way (*bottom-up* analysis) or in a descending way (*top-down* analysis). Dialogue systems typically use the bottom-up analysis. The first type of analysis carried out is lexical (or morphologic), which attempts to detect the individual words and assigns them to lexical categories such as names, verbs, adjectives, etc. One a problem with this kind of analysis is the ambiguity (called *lexical ambiguity*) that occurs when a word (e.g. 'right') can be assigned to several lexical categories (e.g. privilege, right-hand side, agreement). After the lexical analysis has been performed, the next stage is the *syntactic* analysis, although some dialogue systems omit this stage. This analysis aims at building structures that represent the grammatical function of words, which implies a transformation from the lineal sequence of words obtained from the speech recogniser into a hierarchical structure of words. The ambiguity problem also occurs at this level of analysis (now called *syntactic* or *structural ambiguity*) when several hierarchical structures (trees) can be obtained for the input word sequence. The third stage of analysis is called semantic analysis, which attempts to discover the meaning from the syntactic structures obtained in the previous analysis (or that from the individual words found in the lexical analysis). This analysis is fundamental for SDSs since to interact with a user, these systems must understand his sentences. Finally, the last kind of analysis is the pragmatic one, which tries to obtain the meaning of the sentences within a determined context, i.e. what the user really wants to say. The classical example is the sentence 'Do you have the time?', uttered not to obtain a yes/no answer but to know the current time. In some systems there is no clear frontier between these analyses, which are carried out in a more or less combined way.

The module of the dialogue system that implements the NLP must face a diversity of problems. Some of them are classical problems of NLP such as *ellipsis*, which occurs if some words are omitted in sentences if they are implicit in the context. Another problem is the *anaphora*, which occurs when a word makes reference to some word previously mentioned. For example, in the sentence 'I'd like it big' the word 'it' refers to another word that must be found by the system in order to understand the sentence. The third classical problem is ambiguity, mentioned previously. But in addition to facing these problems, the NLP module must also face the limitations of the current ASR technology, which implies recognition errors (insertions, deletions and substitutions of words) that may create ungrammatical sentences. Also, the NLP module must deal with spontaneous speech, which implies that sentences may contain hesitations (e.g., 'uhm . . . '), false starts (i.e. corrections when the user starts to speak, e.g. 'On Mond – Tuesday') and other features that can make the input to the NLP module ungrammatical.

A variety of grammars can be found in the literature to carry out the NLP. For example, phrase-structure grammars are efficient for some applications but are difficult to maintain (e.g. add new rules) and adapt to other applications (Borsley 1996). Context-free grammars are very appropriate to describe well-formed sentences, and have been widely used because very efficient parsing algorithms have been developed (Satta 2000). However, they are not very appropriate when dealing with spontaneous speech, and therefore have not been used in SDSs. Semantic grammars are characterised because their rules are applied considering syntactic and semantic considerations simultaneously, which presents several advantages (Su et al. 2001). On the one hand, the simultaneous application of syntactic and semantic considerations is useful to avoid possible ambiguities, since some hierarchical structures can be discarded if their corresponding semantic interpretations have no meaning. On the other hand, these grammars ignore syntactic aspects that do not affect the semantic content of sentences. Several studies indicate that these grammars are very useful to process natural language in restricted domains. Their main drawback is that they may require a great number of rules for some applications, and thus the sentence analysis can be costly.

2.1.3 Face Localisation and Tracking

Many HCI applications require the localisation of a human face in an image captured by a camera. This task is very important for face tracking since in order to track the face in a sequence of images, the tracking system must start by localising the face. Face localisation is particularly important for MMDSs since it allows to obtain specific information about facial features of the user, as the lips and eyes. The integration (or *fusion*) of acoustic and visual information obtained from the lip movements can notably enhance the performance of acoustic-only ASR systems, which work quite well when processing clean signals but perform poorly when processing low-quality signals due to noise or other factors.

Face localisation can be considered a particular type of the face detection task, simplified to the case when there is only one face in the images. The first step towards localising the face is to detect it, which is a difficult task for several reasons (Yang et al. 2002). First, face images vary due to their relative posture towards the camera. Second, the presence or absence of facial features, for example, a beard, a moustache, etc., cause a great degree of variability, thus making the localisation of the face more difficult. Third, the diversity of facial expressions is also an important factor, since the appearance of a face is directly related to the facial expression of the subject. Fourth, images may present a problem of occlusion, i.e., faces can be partially occluded by other objects. Finally, illumination factors (e.g. spectra, light source features, light intensity, etc.) and camera conditions (sensor response, lenses, etc.) notably affect the appearance of the face.

Several approaches have been proposed to carry out face localisation. This section briefly discusses those based on knowledge, invariant features, pattern matching and colour (Yang et al. 1998, 2002). The knowledge-based approach uses rules that encode human knowledge about what constitutes a face. Usually, the rules capture the relations between facial features.

For example, a face generally has two symmetric eyes, a nose and a mouth, and the relations between these features can be represented by their distances and relative positions. The detection methods first extract the facial features, and then identify candidate faces using the rules. One problem with this approach is the difficulty in representing human knowledge about what constitutes a face. Thus, if the rules are too strict, the method may fail in detecting faces that do not satisfy all rules, while if the rules are too general, there will be many areas in the images that will be considered faces although they are not.

The approach based on invariant features aims to find structural features that exist on faces although the position, point of view or lighting conditions change. The basic assumption is that human beings can easily detect faces in different positions and lighting conditions, thus there must be some properties or features that are invariant against such changes. Several methods have been proposed to first detect the facial features and then infer the presence of a face. For example, features such as eyebrows, eyes, nose and mouth are generally extracted using edge detectors. Based on the extracted features, a statistic model is constructed to describe their relationships and verify the presence of a face. One problem with this approach is that the images can be severely corrupted by lighting conditions, noise and occlusion.

The approach based on pattern matching uses several standard patterns that describe a face as a whole, or as separated features. To detect a face, the approach calculates the correlations between the input image and the patterns. This approach has the advantage of being simple to implement. However, a problem is that it is not very good at dealing with variations of scale, position and shape. In order to provide robustness against these variations, several methods have been proposed based on sub-patterns and deformable patterns (Samal and Iyengar 1995; Sumi and Ohta 1995).

The colour of the human face has been used as an effective feature in many applications, ranging from face detection to tracking of hand movement. Although people have different colour of skin face, several studies indicate that the greatest differences are in terms of intensity rather than in terms of chrominance. Several colour-based approaches have been proposed based on the idea that processing colour is much faster than processing other facial features, and also colour is an invariant feature against changes of orientation under specific lighting conditions. This property allows the movement estimation to be carried out more easily, since only a translation model is required. For example, Schmidt et al. (2000) generated a Gaussian distribution model using a set of skin colour regions obtained from a colour-face database, and compared it with the input image in order to identify candidate faces, i.e. skin regions. The presence or absence of a face in each region is then verified using an eye detector based on a pattern matching process, since there is biological evidence that the eyes play a very important role in the detection of human faces. Once the face is detected, pupils, nostrils and lip corners are used to track its movement in the sequence of images.

Other authors have used histograms, which are colour distributions in the chromatic space (Crowley and Bedrune 1994; Crowley and Berard 1997), since it has been shown that they are stable representation of objects that can be used to differentiate a great number of objects. It has also been demonstrated that the face colour distribution forms clusters in specific points of the plane, as can be observed in Figure 2.1. There are several problems in using colour-based information to locate and track the user face. First, the face colour obtained by a camera can be influenced by many factors, for example, the environmental lighting,

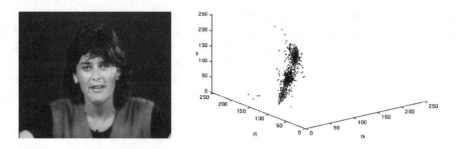

Figure 2.1 Example of a human face and the skin-colour cluster in the RGB space (reproduced by courtesy of Dr J. Yang, Carnegie Mellon University, Interactive Systems Labs, Pittsburgh, PA, USA)

the movement of other objects, etc. Second, different cameras can produce different values for the input colour information, even for the same person and under the same lighting conditions.

2.1.4 Gaze Tracking

The function of human gaze in human-to-human dialogues has been studied for several years. The way speakers and listeners look or avoid their mutual visual contact, the fact of looking towards the speaker or anywhere else, etc., are issues that have been extensively studied taking several parameters into account, for example, age, gender, personal traits, interpersonal relationships, etc. Some interesting behaviour patterns have been found, for example, it has been observed that gaze helps regulate the interaction flow and plays a very important role in ensuring the coordination in the turn taking.

Gaze tracking systems are very useful in a variety of applications. One example is the teleconference, in which it is desirable to track the speaker's face moving freely in a room. But what is more important for the purpose of this book, tracking the user face plays a very important role in MMDSs since it can provide information about the current interest of the user, identified by the point of the screen or other surface at which he is looking. This point may be a specific object in the context of the conversation, so that this information may be used by the system to enhance the ASR or resolve a referring expression. Moreover, gaze tracking allows regulation of the turn taking and provides information about the state of mind of the user, as well as other interpersonal attitudes.

Gaze tracking systems have existed for several years, but were restricted to lab experiments. Recent technological advances have made them robust and cheap enough to be used in dialogue systems. They represent an example of new HCI technologies that are not based on commands, since the user does not execute commands explicitly; instead, the computer observes him passively to provide the appropriate responses or carry out determined actions.

Many research efforts have been devoted to the study of gaze tracking in order to use this information as an input modality, typically to replace the movement of a mouse or some other pointing device. This is of great use, for example, for disabled users as well as for applications that require hand-free interaction. There are gaze tracking systems based on reflected light (Pastoor et al. 1999), electric skin potential (Gips et al. 1993) and contact lenses (Bour 1997). The techniques of the first kind are non-intrusive and are based on the reflection of pupil or cornea, Purkinje[1] images, etc., while techniques of the second and third types are intrusive. Generally, LEDs are used as point light sources since they are safe and cost effective. Most non-intrusive methods require users to make a tedious calibration of parameters (e.g. radius of the cornea, head position, etc.), while others require the head to be located in a very narrow area. Several techniques have been proposed in order to reduce these inconveniences. For example, Shih et al. (2000) presented a method that uses several cameras and multiple light point sources to estimate the sight line without needing to calibrate user-dependent parameters, thus reducing the inconvenience and calibration errors. Ohno et al. (2003) proposed a method that only requires calibration the first time the user interacts with the system; the method uses a stereo camera unit in front of the user to locate and track his eye position if he moves the head.

Several gaze tracking methods can be found in the literature, ranging from intrusive hardware devices mounted on the user's head, to non-intrusive software methods based mainly on software. For example, Stiefelhagen and Yang (1997) presented a non-intrusive method for gaze tracking that estimates the position in the 3D space of the user head by tracking six points of facial features (eyes, nostrils and lip corners) (Figure 2.2). The method locates the user face using information about colour and is able to find and track the points automatically as long as a person appears in the vision range of the camera.

Also, Yang et al. (1998) presented a face tracking system that in addition to using a colour-based model to locate the face, uses a model that estimates the movement of the image and predicts the location of a search window used to locate facial features (eyes, lip-corners and nostrils). The method also uses a camera model to predict and compensate for the camera movements (panning, tilting and zooming), allowing tracking of the face as the user moves freely about a room, walks, jumps, sits down and stands up. Once the user face is located, the eyes are tracked automatically and the obtained images (Figure 2.3) are processed to be provided as input to an ANN that estimates the 'x' and 'y' screen coordinates at which the user is looking.

Figure 2.2 Detected eyes, nostrils and lip corners (reproduced by courtesy of Dr R. Stiefelhagen, Univ. Karlsruhe (TH), Interactive Systems Labs, Germany)

---

[1] When a light beam crosses a surface that separates two media, part of the light is reflected by the surface while the other part is refracted after crossing the surface. The light reflected by the exterior cornea surface is called the first Purkinje image.

Figure 2.3 Sample input images for the neural network (20 × 10 pixel) (reproduced by courtesy of Dr R. Stiefelhagen, Univ. Karlsruhe (TH), Interactive Systems Labs, Germany)

2.1.5 Lip-reading Recognition

Lip-reading plays a very important role in the communication of hearing impaired people, and is also useful for people without any hearing disability when they communicate in a difficult acoustical environment. As commented above, most approaches to ASR are based on acoustic and grammatical information only, but the acoustic part is very sensitive to distortions caused by background noise. Moreover, using this technology is not suitable when several voices are present simultaneously (the so-called 'party effect'). Human beings are able to deal with these distortions taking into account directional, contextual and visual information, mainly concerned with the lip movement. The visual signals influence the perception of speech in face-to-face conversations. Some of these signals provide information that allows semantic information to be extracted from visual-only stimuli, using the so-called *lip-reading* or *speech-reading* techniques. Several experiments show the notable effect of visual cues in the perception of speech signals. Two classic examples are the 'McGurk' and the 'ventriloquist' effects. In the first case, when a spoken 'ga' sound is superimposed on the video of a person uttering the 'ba' sound, most people perceive the speaker as uttering the 'da' sound (McGurk and McDonald 1976). In the second case, listeners hear speech sounds at the location of a visual representation of the speaker face even when the face is partially displaced from the origin of the audio signal (Bertelson and Radeau 1976).

There are several research projects that aim to enhance the quality of life of people with hearing disabilities using lip-reading techniques. For example, SYNFACE (Karlsson et al. 2003) attempts to allow them to use an ordinary telephone by making the incoming speech signal govern the lip movements of a talking head that provides support to the user. The basis is that there are three reasons why vision provides people with a great advantage in perceiving speech: it helps locate the sound source, contains segmental information that supplements the audio information, and provides complementary information about the articulation place. The last reason is a consequence of the partial or complete vision of the articulatory organs, as the tongue, teeth and lips. The information regarding the articulation place can help disambiguate, for example, the consonants /b/ and /d/, or /m/ and /n/, which are very easily confused if only the acoustic information is considered. Additionally, the movements of chin and lower face muscles are correlated with the produced sounds, and it has been demonstrated that seeing them enhances the human perception of speech. The movements of the oral cavity associated with the phonemes that are visually similar constitute what is called a *viseme*, which is the smallest visually distinguishable unit of speech. Generally, visemes are defined by a many-to-one mapping from the set of phonemes, which reflects the fact that many phonemes (e.g. /p/ and /b/) are practically indistinguishable if only the visual information is considered.

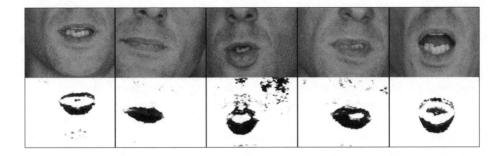

Figure 2.4 Images of ROI for lip-reading (top row) and lip detection (bottom row) (reproduced by permission of Daubias and Deléglise 2002)

Lip-reading recognition introduces new challenges to the MMDS discussed so far. First, the visual front-end requires extraction of visual information from the user face, which relies on the face localisation task commented above. Second, it requires extraction of the visual characteristic features. In most lip-reading systems, the ROI (region of interest) is a square that contains pixels of the speaker's mouth region (Figure 2.4), which is typically applied in a normalisation process (in terms of scale, rotation, lighting compensation, etc.). The ROI usually also includes other areas of the lower part of the face (such as chin, cheeks, teeth or nostrils) since it has been widely accepted that the information of lip movement is not enough for lip-reading.

The literature describes various lip-reading recognition techniques (Duchnowski et al. 1995; Basu et al. 1998; Revéret and Benoît 1998; Daubias and Deléglise 2002). For example, Duchnowski et al. (1995) used a lip locator to allow user free-movement interaction using speech in a possibly noisy room. The lip locator uses two ANNs, the first to obtain an approximate estimate of the mouth position, and the second to locate the two lip corners from the estimation provided by the first network. The mouth position is the input to an AVASR (Audio-Visual Automatic Speech Recognition) system (studied in Section 3.2.2.1) based on two MS-TDNN (Multi-State Time Delay Neural Networks) to recognise German letters by the integration of visual and spoken information chunks at the phoneme/viseme level.

Daubias and Deléglise (2002) used a shape-appearance model to extract visual information from the user face. Concretely, to obtain the shape model they used deformations of a 2D polygonal approximation obtained from a speaker corpus, while to obtain the appearance model they used an ANN, more specifically a multilayer perceptron with one input layer, one hidden layer and three output units. For each input block, the output units indicate the probabilities of 'skin', 'lips' and 'inner-mouth' classes. For the visual recognition they used three parameters (and their first and second derivatives): inner-lip horizontal width, inner-lip vertical height and inner mouth area. These parameters were calculated automatically using the inner-lip contour from the polygonal lip-shape approximation. The visual measures obtained from the model were integrated with acoustic measures in the AVASR system.

2.1.6 Gesture Recognition

The study of body gestures is a research field of great interest nowadays, promoted partly by the recent advances in the field of virtual reality. Gestures constitute an information modality that is very integrated in the human communication ability, allowing for the specification of objects and operations very efficiently. The current recognition technology can be used in many applications, ranging from advanced human-computer interfaces to new interactive multimedia systems. In the context of MMDSs, this technology is concerned with the capture and analysis of movements of the body (or parts of the body) in the 3D space to express or give emphasis to ideas or concepts. Some authors (e.g. Martin et al. 1998) consider that the movements made in the 2D space using a mouse or a pen on a touch-sensitive screen can also be considered part of this technology.

Dialogue systems must face different kinds of problem when recognising gestures. The acquisition itself may be a problem more or less complicated depending on the device used. The problem can be minimal for devices such as pens and touch-sensitive screens but may be very important for other devices, for example, magnetic trackers, cyber-gloves or cameras for artificial vision. The reason is that pen and touch-screen devices generate significant information only, whereas devices that acquire information continuously (e.g. magnetic trackers, cameras for artificial vision, etc.) generate both significant and insignificant information. Thus it is necessary to detect the significant data in the streams of information provided by these devices.

Gesture recognition can be carried out using intrusive (e.g. data gloves) or non-intrusive devices (e.g. cameras for artificial vision) depending on the application requirements. Data gloves are typically used in virtual reality applications to recognise 3D gestures, while 2D gestures are typically recognised using touch-sensitive screens or tablet digitisers. For example, Chu et al. (1997) proposed a three-stage method to recognise 3D gestures using a data glove. The first stage carries out a characteristic abstraction that obtains diverse data from the glove, for example, the position of fingers and the orientation and movement of the hand. The second stage builds data structures called *gestlets* to store information about the posture, orientation and movement from the glove data. Finally, the third stage analyses the data in the gestlets to determine the type of gesture made by the user.

In other applications intrusive devices are not acceptable and then non-intrusive devices must be used. This type of gesture recognition requires locating with a camera the subject that will carry out the gestures, locate his hands and arms as well, and finally classify his gestures (Krumm et al. 2000). Visual recognition of hand gestures can help increase the ease and naturalness desired in MMDSs, which has motivated a great amount of research. There are significant differences in the recognition approaches depending on whether 3D models of hands, or image appearance models of hands are used. 3D models allow modelling gestures in a more elaborated way but present computational limitations that make them unsuitable when considering the real-time requirements of MMDSs. Models based on image appearance have low computational requirements and work well for limited applications, although seem to lack something in terms of generality. The choice of the most suitable acquisition device for a dialogue system also depends on the platform in which the system is set up. Pens are usually preferred for mobile platforms such as PDAs, since they provide fast and accurate recognition. Cyber-gloves or magnetic trackers can also be efficient and accurate but need to be carried or worn by the users. Gesture recognition approaches based on artificial vision offer a completely free interaction and are flexible enough to be used in all platforms, except in the smallest mobile devices.

2.1.6.1 Classifications of Gestures

Using gesture recognition in MMDSs requires an understanding of how people make gestures. On occasions gestures are used as the only communication modality but in many others are used in combination with other modalities. The literature shows different gesture classifications. In a first approximation, it is possible to distinguish *gross-grain* (movement of the whole body), *medium-grain* (e.g. gestures made with arms, legs, etc.) and *fine-grain* gestures (gestures made with fingers). Other possible classification divides them into *static* (considered isolated in individual images) and *dynamic* gestures (considering motion trajectory).

McNeill (1992) studied gestures from the psycholinguistic point of view and considered them a critical step between the linguistic and conceptual abilities of human beings. People tend to express using gestures what they cannot express via speech. Thus, gestures are sometimes considered as a means to express thoughts or ways of understanding the events of the world. Gestures combined with speech have been found to be a very effective means of communication to deal with the complexity of the visual space.

According to Scherer and Ekman (1982) and McNeill (1992) gestures can be classified into five categories: symbolic, deictic, iconic, metaphoric and beat or rhythmical. *Symbolic* gestures are used to convey some meaning (e.g. move the head to affirm or negate). *Deictic* gestures are used to point to objects or places (e.g. saying 'put that there' while pointing to an object and a location). *Iconic* gestures are typically used to describe objects, relations or actions visually. *Metaphoric* gestures are used to describe abstract ideas and events. And finally, *beat* or *rhythmical* gestures are made synchronously with speaking (e.g. with hands and arms). Iconic and metaphoric gestures can be either complementary or redundant to the oral channel and can provide additional information, robustness or emphasis to what is being said. Symbolic gestures can be interpreted without a reference context; however, deictic, iconic and metaphoric gestures need such a context to be interpreted correctly, which may be provided by other input modalities used simultaneously.

Kettebekov and Sharma (2000, 2001) proposed a testbed framework called iMAP (interactive map) to carry out studies about hand gestures in combination with speech in a multimodal interaction. They focused on a typical scenario in which a person makes gestures in front of a map while he narrates the weather conditions (Figure 2.5a). The goal of their study was to find techniques that could be applied to the design of recognition systems for free hand gestures. In the studied domain, the authors classified hand gestures into two groups: *transitive* deixis and *intransitive* deixis. The first kind includes the gestures used to make reference to concrete objects without indicating a change in their physical location. Here, the *transitive* feature means these gestures must be completed with verbal information. These are the *deictic* gestures commented before, but the authors considered three types of such gestures: nominal, spatial and iconic. The *nominal* deixis includes gestures used to select objects by referring to them using a name or a pronoun. The *spatial* deixis includes gestures that are generally completed with adverbs of the form 'here', 'there', 'under', etc. Finally, the *iconic* deixis is a hybrid of the two previous types since it is not only used to make

(a)   (b)

Figure 2.5 Two analogous domains for the study of natural gestures: (a) The weather narration and (b) The computerised map interface (iMAP) (reproduced by permission of Kettebekov and Sharma 2001)

reference to determined objects, but also to specify their attributes in a spatial continuum (e.g. their shape).

The *intransitive* deixis includes gestures concerned with a spatial displacement. This type of gesture is analogous to the verbs of movement (e.g. move, enter, etc.). Contrary to the transitive deixis (which requires reference to a concrete object to complete its meaning), intransitive deixis requires a continuum spatial-temporal relation (i.e. a direction).

2.1.6.2 Approaches to Gesture Recognition

Many methods have been proposed for 2D and 3D gesture recognition, mainly based on templates (Hu et al. 2003), ANNs (Su et al. 1998), HMMs (Pavlovic et al. 1997), Bayesian Networks (Serrah and Gong 2000), Principal Component Analysis (Lamar et al. 1999) and artificial vision techniques (Hu et al. 2003). Some hybrid approaches have also been proposed, for example, Johnston et al. (1997) combined ANNs and HMMs to recognise 2D gestures made on a map. Template-based approaches compare a given input template against one or more prototypical templates of each expected gesture. For example, the pedestrian navigation MATCH system (Johnston et al. 2002) used a variant of a template-based gesture recognition algorithm trained on a corpus of sample gestures, including lines, arrows, areas, points and question marks. The Jeanie system (Vo and Wood 1996) developed for the appointment scheduling domain, used Time Delay Neural Networks (TDNNs) to recognise 2D gestures (concretely basic shapes such as lines, arcs, arrows, circles, etc.) considered decomposed into sequences of strokes.[2]

The approach based on HMMs considers that gestures made by the same or different users likely differ in shape and temporal properties, which makes it very difficult to find manually a general description of their spatio-temporal patterns. The recognition is modelled as a stochastic process similar to the one used in ASR. For example, Lee et al. (1996) presented

---

[2] A stroke can be defined as the set of points between a pen-down and the next pen-up. A word consists of a sequence of one or more strokes.

a system based on HMMs that, in addition to recognising gestures, was able to learn new gestures on-line. Moreover, the system could update interactively its gesture models with each gesture it recognises. It was connected to a cyber-glove to recognise the sign language alphabet.

2.1.7 Handwriting Recognition

Since keyboards are becoming more difficult to use as computers are becoming smaller, the handwriting recognition technology makes it easier to input text into the computer by writing instead of typing, using a device such as a digital pen, a stylus or simply moving the mouse. The computer transforms the handwriting into written text and inserts it where specified, providing visual feedback that is concrete and easy to edit. Moreover, the combination of handwriting and speech synthesis is very useful since the user can obtain acoustic feedback of the written characters that have been recognised by the system. The technology still is not very robust and the interaction speed is lower than with speech. In the case of MMDSs, handwriting is particularly useful to input data in mobile systems and noisy environments.

Handwriting recognition has many features in common with ASR, for example, user independence/dependence, existence of different possible units (characters, digits, words or sentences), different writing styles (e.g. printed versus cursive), vocabulary size, etc. The literature shows several classifications of recognition methods, that initially can be classified into two types: *off-line* and *on-line*. In the first type, also known as optical character reading (OCR), only spatial information is available since writing is provided as an image, therefore no information is available regarding the temporal sequence of recognition units. In the second type, which is the one employed in MMDSs, spatial-temporal information is available and recognition is made taking into account the writing device movement, represented by a temporal coordinate sequence. To carry out this recognition type a tool is required, e.g. a pen or stylus with which the user writes on a flat device (e.g. a touch-sensitive screen) and software that interprets the movements, transforming the trajectories into written text. Some recognition difficulties are concerned with the variation in the number of strokes, the connection of the strokes, the variation in the order of making the strokes, and the variation of their shape (for instance, a Japanese Kanji can constitute up to 30 strokes, although it is usually written connecting several strokes for simplicity).

Handwriting recognition methods can also be classified into *analytic* and *holistic* (Duneau and Dorizzi 1996). The former makes hypothetical segmentations before the recognition process begins. For example, it aims to identify initially the individual characters (using a technique for character recognition) and from the hypotheses obtained it builds hypotheses about words. By contrast, the holistic approach considers the input as an indivisible whole, so that no initial segmentation is carried out. The main disadvantage of the latter is that it is only appropriate in applications of small vocabulary size.

The quality and usability of handwriting technology depend on three parameters: vocabulary size, writing style and training. Each handwriting recogniser uses a vocabulary that contains reference models of the possible recognition units (e.g. characters, digits or words). For example, the handwriting recogniser of the pedestrian navigation MATCH system (Johnston et al. 2002) uses a vocabulary with 285 words, including attributes of restaurants (e.g. 'cheap', 'Chinese', etc.) as well as zones and points of interest (e.g. 'Soho', 'empire', 'state', 'building', etc.).

The second factor is the writing style. If users are asked to use a specific writing style (e.g. capitals) the input variability will be smaller and thus the recognition accuracy will increase. However, this kind of restriction tends to reduce usability. The third factor is the training: if the recogniser allows the user to train the recognition models in his own writing style, recognition rates will be improved.

A variety of devices can be used in handwriting recognition, for example, tablet digitisers, touch-sensitive screens, LCD tablets, etc. An important feature of these devices is how much the device resembles the feel of a paper and a pencil. Some known problems are concerned with the delays in the immediate visual feedback and in the sampling of the movement of the stylus, digital pen, etc.

The recognition methods for character units can be classified into three categories. The first is based on a *structural analysis* in which each character is defined in terms of a stroke structure. The second type is based on *statistic methods* that are used to compare diverse features of the characters (for example, their initial and final positions of strokes) with previously created patterns. The third method is based on explicit information about *trajectories*, in which the temporal evolution of the pen coordinates plays a very important role. Each method has advantages and drawbacks. For example, the first two have problems related with the stroke connection, while the third has problems if the same character is written in different order. There are also hybrid algorithms based on the combination of the three algorithms previously described.

Handwriting recognition of sentences is usually based on that of words, including additional syntactic information by means of statistical language models, similar to those used for ASR, to enhance recognition accuracy. For example, Srihari and Baltus (1993) used a trigram and achieved an enhancement of 15% (from 80% to 95%) in a system with a vocabulary containing 21,000 words.

Implementation of handwriting recognition systems is mainly based on ANNs and HMMs. For example, the Jeanie system (Vo and Wood 1996) developed for the appointment scheduling domain used Multi-State Time Delay Neural Networks (MS-TDNN) for handwriting recognition, while Yasuda et al. (2000) presented an algorithm to learn and recognise on-line Japanese written characters (Kanji, Katakana and Hiragana) using HMMs.

2.2 Multimodal Processing

The multimodal processing carried out by the conceptual MMDS shown in Figure 1.1 relies on several modules (multimodal data fusion, multimodal data storage, dialogue manager, task, database and response generation) which are studied briefly in the following sections.

2.2.1 Multimodal Data Fusion

One of the most important challenges in the development of MMDSs is concerned with the integration of information chunks from the different input modalities. Data fusion refers to the combination of these information chunks to form new information chunks so that the dialogue system can create a comprehensive representation of the communicated goals and actions of the user. For example, the pedestrian navigation system presented by Wasinger et al. (2003) uses speech and gestures input in order to enhance speech recognition in different environments (e.g. noisy, mobile, indoors, outdoors, etc.). The system, set up in a PDA, allows user interaction via icons on the display as well as real-world pointing to objects. For example, the user can enter 'What is that?' or 'Describe that landmark to me'. The fusion in this system also facilitates the resolution of referring expressions, for example, the user may point to the 'Mensa' building while he says, 'Tell me about the building over there.'

In order to carry out the data fusion, the designer of the dialogue system must address several issues, for example, the criteria for fusing the inputs, the abstraction level at which the inputs are fused, and what to do if the information provided by different recognisers is contradictory. A typical criterion for fusing inputs is considering time restrictions (e.g. fusion of simultaneous or near in time speech and gesture input). However, when using this approach it must be taken into account that the input devices take different times to analyse the stimuli captured from the user and generate an output, and thus the response times may be relative. For example, a speech recogniser usually takes more time to output a recognised sentence than a touch-sensitive screen to provide the screen coordinates of a point gesture made with a pen. Therefore, the multimodal data fusion module may receive the information captured from the user in the wrong order (e.g. it may obtain the screen coordinates first although the pointing gesture may have been made after uttering a sentence). Using time as fusion criterion, it is usual to assign time stamps to the information generated by the devices. Taking into account the time stamps, the fusion module must decide whether information chunks must be considered to correspond to the current input or to different inputs.

2.2.1.1 Fusion Levels

In order to effectively use the information provided by the input modules studied in the previous sections, MMDSs employ techniques to make the fusion of the different information chunks provided by these devices. The literature shows different levels at which the fusion process can be carried out, ranging from the signal level to the semantic level. The fusion at the *signal level* (also called lexical fusion) represents a binding of hardware primitives to software events; only temporal issues (e.g. data synchronisation) are involved without taking into consideration any kind of high-level interpretation. This fusion type is usually employed for audiovisual speech recognition, which combines acoustic and lip movement signals. Several technologies have been proposed to carry out this type of fusion. For example, Hershey and Movellan (1999) used ANNs trained separately for acoustic and visual recognition, while Su and Silsbee (1996), McAllister et al. (1997), Plumbley (1991) and Deco and Obradovic (1996) used HMMs and Fast Fourier Transformations (FFTs) to analyse the voice spectrum and find the correlation with the lip form. To mention a sample application, Prodanov and Drygajlo (2003) used Bayesian networks for the fusion of multimodal signals in an autonomous tour-guide robot developed for the Swiss National Exhibition 2002. In this application, the networks were used to combine the output of a speech recogniser set up in a mobile robot with signals provided by the robot laser scanner, which are insensitive to acoustic noise. The data fusion meant the robot could discover the user intention when attending the next exhibit presentation. The main disadvantage of the fusion at the signal level is that it does not scale up well, requires a large amount of training data and has a high computational cost.

The fusion at the *semantic level* deals with the combination of multimodal information chunks interpreted separately and then combined after considering their meanings. This type of fusion is usually carried out in two stages. In the first, the events of the diverse modalities are combined in a low-level interpretation module, whereas in the second the result of this combination is sent to a higher-level module that extracts its meaning. The semantic fusion presents several advantages in comparison with the signal level fusion. First, the recognisers for each modality are trained separately and thus can be integrated without retraining. Semantic fusion is also simpler since the integration of the modalities does not require additional parameters to those used for the individual recognisers, which allows scaling up well to include new modalities. Semantic fusion is mostly applied to modalities that differ in the time scale characteristics of their features. Since time plays a very important role in data fusion, all the chunks include time stamps and the fusion is based on some time neighbourhood condition. For example, this kind of fusion was used in the TeleMorph project (Almeida et al. 2002) to combine the input provided by speech and pointing. The temporal relationship between these channels was obtained by considering all input contents within a reasonable time window. The length of this time window had a default value of one second and was a variable parameter that could be adjusted dynamically according to the dialogue context.

The type of fusion chosen for a given application depends not only on the input modalities but also on the particular task carried out by the dialogue system and the type of users. Xiao et al. (2002) indicated that children and adults adopt different multimodal integration behaviour. Therefore, a comprehensive analysis of experimental data may help gather insights about the integration pattern, thus leading to the choice of the best fusion approach. Section 3.2.2 discusses methods of integrating multimodal data and studies the integration of acoustic and visual information, as well as that of speech and gestures.

2.2.2 Multimodal Data Storage

The various input modules transform the information generated by the user into an internal representation that allows the fusion of the information chunks. The multimodal data storage module gathers this information properly represented in a specific format, thus making it available for the dialogue manager to resolve referring expressions and access the dialogue history (McKevitt 2003). Typically, the internal representation allows storage of not only the user information but also about the modules that produce the information, confidence scores, creation time of the information chunks, etc. The literature shows a great variety of methods to represent multimodal data obtained from the user, which are also typically used to communicate the diverse components of a multimodal system. This section briefly describes four of these: melting pots, frames, XML and typed feature structures.

2.2.2.1 Melting Pots

Nigay and Coutaz (1995) proposed the so-called *melting pots* to encapsulate multimodal information with time stamps, as can be observed in Figure 2.6. The lower part of the figure shows the two melting pots obtained when the user says 'Flights from Chicago to this city' at time T_i and points to 'New York' at time T_{i+1} (the melting pot on the left is generated by pointing while the one on the right by the spoken sentence). The fusion process combines both melting pots to obtain a new one with the departure and arrival cities specified. Note

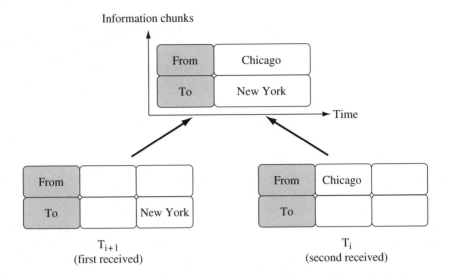

Figure 2.6 Fusion of two melting pots (adapted from Nigay and Coutaz 1995)

that the melting pot on the left was received first by the fusion module given that the response time of the pointing device (mouse) is smaller than that of the speech recogniser.

The fusion of melting pots is carried out following three criteria: complementarity of melting pots, time and context. According to these criteria, Nigay and Coutaz considered three types of fusion: *microtemporal*, *macrotemporal* and *contextual*. The microtemporal fusion combines information chunks provided in parallel. This type of fusion is carried out when the information chunks contained in the melting pots are complementary and overlap in time.

The macrotemporal fusion combines information chunks related and generated either sequentially, in parallel (but processed sequentially by the system), or even delayed due to insufficient processing power of the system (e.g. processing speech takes more time than processing the pressing of mouse keys). This type of fusion is carried out when the information chunks are complementary and the time intervals of the melting pots do not overlap in time but belong to the same analysis window.

Finally, the contextual fusion combines information chunks without considering time restrictions. It combines a new melting pot m with the melting pots M in the current context if m is complementary with any of the melting pots in M, which are ordered in terms of generation time.

2.2.2.2 Frames

Another method of representing multimodal inputs is based on the so-called *frames*. A frame can be defined as an abstract representation of real-world entities and their relationships. Each entity if constituted of a set of attributes (called *slots*) that are filled with the data extracted from the user input (Allen 1995). Figure 2.7 shows an example of a frame representing the 'person' entity and an instance of this frame.

```
[PERSON] →                              [PERSON] →
  [SUPERTYPE] → (mammalian)               [SUPERTYPE] → (mammalian)
  [NAME] → ( )                            [NAME] → (Peter)
  [AGE] → ( )                             [AGE] → (27)
  [HEIGHT] → ( )                          [HEIGHT] → (1.68)
  [GENDER] → ( )                          [GENDER] → (male)
         (a)                                     (b)
```

Figure 2.7   (a) Frame representing the 'Person' entity. (b) Instance of the 'Person' frame

Frames can be classified into three categories: *input*, *integration* and *output*. The information chunks provided by each input device are interpreted separately and transformed into *input* frames that may contain partial information represented by means of empty slots. During the fusion, the input frames are combined by obtaining *integrated* frames whose slots may be filled by the combination of the frames. The filling of the slots of integration frames can be carried out incrementally as the system obtains data from the user. They can also be filled by means of heritage relationships between frames; for instance, in Figure 2.7(b) 'Peter' fulfils the mammalian feature because he is in an instance of the 'Person' frame. Also, frames may have specific functions to fill some slots considering the value of other slots. Some slots can be obligatory (i.e. must contain a value) while others may be optional (i.e. can have or not have a value). Generally, some slots have associated confidence scores that represent the system confidence on the data they contain. Some slots may have filling restrictions (e.g. a slot referring to the age of a person always must contain a positive number). *Output* frames are used by modules that take part in the generation of the system output. For example, the dialogue manager can use a frame to represent its next action (e.g. provide some information to the user) and send it to the response generation module (Section 2.2.6), which will transform it appropriately to transmit the information to the user through one or more output modalities.

Frames have been used in several applications. In the field of dialogue systems, they have been used, for example, in the Chameleon platform, a general suite of tools in the form of a software and hardware platform that can be adapted to conduct intelligent multimodal interfaces in several application domains (Brøndsted et al. 2001). Just to mention some dialogue systems, frames have been used in the setting up of Jeanie (Vo and Wood 1996), Olga (Beskow and McGlashan 1997) and Rea (Cassell et al. 2000). The Jeanie system allows input interaction using speech, gestures made with a pen on a touch-sensitive screen and handwriting, while the response is provided using only speech. The input information generated by the user is captured by a separate recogniser for each modality and transformed into a frame representation. All the frames are merged by the data fusion module of the system in order to obtain the combined interpretation. The frames contain slots for a variety of specific information chunks, such as the action to carry out, or the date and time for meetings. The concepts identified by the system's NLP module determine the slots that must be filled as well as the confidence scores that must be assigned to the slots. Frames filled partially are merged to obtain their combined interpretation using a domain-independent algorithm, which allows a high-level processing of the information provided by each input modality. This way, adding a new interaction modality to the system only requires creating a new module to convert the output information generated by the corresponding recogniser into (partial) filled frames, that can be merged with the other frames. Some systems use a slot to identify the module that produces each frame (e.g. speech recogniser, NLP, etc.), another slot to include the input data from the corresponding recogniser (e.g. a sentence, pointing coordinates, etc.), and one or two additional slots to include the start and termination times of the frame processing.

2.2.2.3 XML-based Languages

XML (Extensible Markup Language)[3] provides a natural way to represent information, facilitating modularity and extensibility. A variety of XML-based languages have been used in several dialogue systems. For example, MATCH (Johnston et al. 2002) uses messages encoded in XML, SmartKom (Wahlster et al. 2001) uses M3L (Multimodal Markup Language), MIAMM (Reithinger et al. 2002) uses MMIL (Multimodal Interface Language), MUST (Almeida et al. 2002) uses MXML, etc. Interacting with the SmartKom system, if the user points on a map to the location of the cinema he wants to visit, the system generates a M3L representation as shown in Figure 2.8, in which there are two hypotheses ('dynId30' and 'dynId28') for the object referenced by the pointing gesture. The first object is the one with the highest priority (1) which corresponds with the 'Europe' theatre located in the specified coordinates.

```
<gestureAnalysis>
[...]
  <type> tarrying </type>
  <referencedObjects>
   <object>
   <displayObject>
     <contentReference> dynId30 </contentReference>
   </displayObject>
   <priority> 1 </priority>
   </object>
   <object>
   <displayObject>
     <contentReference> dynId28 </contentReference>
   </displayObject>
   <priority> 2 </priority>
   </object>
  </referencedObjects>
  <representationContent>
  [...]
   <movieTheater structID=dynId30>
   <entityKey> cinema_17a </entityKey>
   <name> Europa </name>
    <geoCoordinate>
     <x> 225 </x> <y> 230 </y>
    </geoCoordinate>
  [...]
  </gestureAnalysis>
```

Figure 2.8   M3L data representation in the SmartKom system (reproduced by permission of Wahlster et al. 2001)

---

[3] http://www.w3.org/XML/

2.2.2.4 Typed Feature Structures

Typed feature structures (TFSs) (Carpenter 1992) are used in MMDS as a very compact means of combining information chunks across modalities and time. According to Alexandersson and Becker (2001), a set of typed feature structures, S_{TFS}, can be recursively defined as follows:

$$S_{TFS} = \{\langle\, t, [a_1 : f_1, \ldots, a_n : f_n]\rangle\} \tag{2.8}$$

where t is the type, $a_1, \ldots, a_n \in A$, $n > 0$, are the attributes that belong to a finite set A, V is a possibly infinite set of values, and $f_1, \ldots, f_n \in V = A \cup S_{TFS}$ are the features. A TFS is usually denoted graphically as follows:

$$\begin{bmatrix} t \\ a_1 \;:\; f_1 \\ \ldots \\ a_n \;:\; f_n \end{bmatrix}$$

For example, the sentence 'I'd like to go to the movies tonight' could be represented as follows (Alexandersson and Becker 2003):

$$\begin{bmatrix} \ldots entertainment : \begin{bmatrix} PERFORMANCE \\ beginTime : tonight \\ cinema : [CINEMA] \end{bmatrix} \end{bmatrix}$$

TFSs facilitate the specification of partial information chunks. For example, sentences or gestures that partially specify the user intention can be represented using sub-specified TFSs in which some features are not instantiated. Thus, the dialogue system must ask the user for the missing information in order to fill in the TFSs and discover his intention. Several authors have used TFSs to implement a variety of dialogue systems. For example, Johnston et al. (1997) used these structures to set up the QuickSet system (a multimodal pen/voice system that enables users to set up and control distributed interactive simulations). Yang et al. (1999) used TFSs to set up the Smart Sight to provide assistance to tourists; concretely, TFSs were used to combine the (partial) information provided by the different recognisers, and also to express the system goals.

In addition to representing multimodal data, Pfleger et al. (2002) and Alexandersson and Becker (2003) used TFSs to combine information chunks using *unification* and *overlay* operations (studied in Section 5.4.4) and obtain other TFSs enriched with contextual information. As we will discuss in Section 5.4.4, the unification operation succeeds if the information chunks to be combined do not have conflicting parts, but fails if conflicting parts (e.g. different types) are found in the TFSs. In case of conflict, the overlay operation can be used to combine the information chunks, basically replacing the old data with the new data.

Holzapfel et al. (2002) used multidimensional TFSs to allow the tagging of each information chunk with recognition confidence scores, information about the modality used to provide the information, etc. The authors used a dimension to store emotional information which allowed it to combine the recognised words with information about the user's emotional state when uttering the words.

2.2.3 Dialogue Management

The dialogue manager is the module that implements the 'intelligent' behaviour of the system. It receives the internal representation provided by the multimodal data fusion module and decides the next action the system must carry out. Typical actions are: (1) to query the database module; (2) to generate a prompt to ask the user for additional data; (3) to generate a confirmation prompt to confirm unreliable data obtained from the user; and (4) to provide help to the user, etc. The decision about the next system's action is made by taking into account the system task module (discussed in Section 2.2.4) as well as a dialogue model and interaction strategies (studied in Sections 5.2 and 5.3, respectively) designed in advance to provide the user with as natural, efficient and comfortable an interaction as possible. The dialogue model is typically built from the analysis of a corpus of human-to-human dialogues collected in the application domain for which the system is designed. Generally, this model is refined and validated using the WOz technique to take into account the specific features of a typical human-to-system interaction (studied in Section 3.2.1.1). Given the current limitations of the diverse recognition technologies involved in the setting up of MMDSs, a variety of dialogue management strategies are available in order to avoid, as long as possible, the effects of such limitations in the subsequent dialogue. These strategies include interaction strategies (e.g. system-directed, user-directed or mixed) and confirmation strategies (implicit, explicit, ask again, etc.).

As commented above, the dialogue management carried out by SDSs is based on the semantic representation obtained from the sentence recognition; in other words, it is based on just one input modality. On the contrary, MMDSs use several input modalities (e.g. speech, gesture, gaze, lip movement, etc.) so that, in these systems, the decisions made by the dialogue manager about its next turn in the dialogue (also called *dialogue move*) must be based on the semantic interpretations obtained from several input modalities, which can provide information chunks simultaneously to the system. The use of several input/output modalities has important implications in the dialogue management, not considered in the case of SDSs, which are discussed in Section 5.4. Additionally, the use of several languages also has implications in the dialogue management that are discussed in Section 5.5.

2.2.4 Task Module

The task module stores a representation of the service provided by the dialogue system, such as train or airplane travel information, directory assistance, weather forecasts, tourist information, etc. The reason for using this module is to have a clear separation between dialogue knowledge and task knowledge in order to facilitate the modelling of dialogue strategies and the maintenance of the system. In order to facilitate the setting up of new systems, the so-called *task-independency* feature aims at developing systems in the form of *platforms* that can easily be adapted to different tasks. Implementing such a platform requires that its modules for ASR, NLU, dialogue management, NLG, etc. are task-independent.

Task-independency has been addressed by several researchers who have focused on different modules of the *platform*. For example, Komatani et al. (2001) focused on the ASR and NLU modules, and proposed a portable platform to develop SDSs to be applied to various information query tasks. This platform automatically generates domain-independent lexicon and grammar rules extracted from a domain database. The portability of the language model is carried out by adopting a simple key-phrase spotting strategy. Using this strategy with the generated grammar rules and a word bigram model trained with dialogue corpora of similar domains, their system carries out flexible speech understanding on a variety of utterances. The semantic analysis is achieved by filling query slots with keywords, and translating key-phrases into them.

Hardy et al. (2004) focused on the platform's dialogue manager and response generation modules. To carry out the dialogue management they used a *task identifier* and a *dialogue act classifier* that were domain-independent. Both identifiers were trained by using a corpus of over 500 conversations from a British call centre, which were previously recorded, transcribed and annotated (the annotation of dialogue corpus will be studied in Section 5.1). Using a vector-based approach proposed by Chu-Carroll and Carpenter (1999), the task identifier represented tasks as vectors of terms built on the utterances requesting them. The dialogue act classifier was trained by using speech acts (Grosz and Sidner 1986) such as 'Accept', 'Reject', 'Non-understanding', 'Opening', 'Closing', 'Backchannel', and 'Expression', obtained from the same corpus. Its goal was to identify the caller's utterance as one or more domain-independent speech acts. For the response generation they stored all language- and domain-specific information in separate files, and used a template-based system, easily modified and extended, to construct utterances according to the dialogue manager's specification of one or more speech acts.

Niimi et al. (2000) also focused on the dialogue manager, and presented a task-independent dialogue control scheme that was applied to the SDSKIT-3 system, which was initially designed for the tourist information service. Their simulation experiments proved that the scheme could be adapted to another service (guide for personal computer buyers).

2.2.5 Database Module

The dialogue system's database shown in Figure 1.1 contains the data the user may be interested in. For example, in the case of a system designed for the ATIS (Air Travel Information Service) domain, the database would contain data about departure/arrival times and gates, travel price and duration, etc. The database, although shown conceptually as just one module in Figure 1.1, can be comprised of several modules, and may contain several types of information. For example, the MATCH system (Johnston et al. 2002) uses two databases, one for subway and the other for restaurant data.

The database can be local (e.g. be located in the same computer running the dialogue system or in a physically near one) or remote (i.e. located in a distant computer, typically accessible through the Internet). For example, the Jupiter system designed to provide weather forecast information (Glass et al. 2000), queries several Internet URLs to find the weather forecasts. In one of these URLs it finds weather and temperature forecasts for the following four days for a wide number of cities all around the world, while in other URL it finds more detailed information, but limited to cities in the United States. The obtained information is processed semantically and stored in a local database as semantic frames.

The database may also contain information about known users (e.g. customers) that may be useful to identify the current user and personalise the interaction (e.g. using the user's favourite system's voice, including the customer's name in the system prompts, etc.). This is the case, for example, of the Amitiés dialogue system designed to provide bank account information (Hardy et al. 2004).

Some kind of emotional information can also be stored in the database, in the form of pre-recorded user emotions expressed through different modalities. This makes it possible to have e.g. an 'anger' gallery with anger facial expressions and anger speech utterances. The *emotional database* can also be used to recognise emotions by comparing the user actual state with the ones stored in the database, and selecting the closest one according to a distance measure. For example, Tsang-Long et al. (2004) used this approach to recognise four user emotions (happiness, boredom, anger and sadness) taking into account their verbal expressions.

2.2.6 Response Generation

The system response decided by the dialogue manager is represented in a specific internal format (e.g. a XML-based language) that is sent to the last module of the multimodal processing component: the response generator. This module carries out the *fission* operation (opposite to the fusion process discussed in Section 2.2.1), consisting of choosing the output to be produced through each output modality and coordinating the output across the modalities. This task is very important for MMDSs in order to make the information coherently presented to the user. For example, lip movements and facial expressions of animated agents (studied in Section 3.2.4) must be coordinated with the speech output. Moreover, if an animated agent refers to an object shown graphically on the screen, the reference and the presentation of the object must be synchronised. The reference to the object can be carried out using a variety of graphic modalities (e.g. a deictic gesture or the gaze of the agent, highlighting the object, etc.), or even can be cross-modal, using a spoken message (e.g. 'The image on the left corner of the screen …'). In this latter case, the reference requires that the modality that makes the reference can access the internal representation of the screen contents; for example, the SmartKom system uses a 3ML representation for these contents, which allows the visual objects be part of the discourse representation and thus be referenced multimodally.

The presentation mode affects the user perception, comfort and preferences when interacting with a dialogue system, as shown by several experiments. For example, Elting et al. (2002) studied the effects of screen size, device type and presentation style on cognitive load, information processing and user acceptance. They note that when the system has to present information that should be remembered (e.g. a city name), the most suitable presentation modality is the one that overloads the user cognitive load (e.g. speech and graphics). In contrast, when the system has to provide other types of information that are not so important, the most appealing/accepted modality must be used (e.g. graphics, text or speech). This study was applied to the TeleMorph project (Solon et al. 2004) to determine the most suitable modality combinations for presenting information.

Dialogue systems based on speech and pen interaction are specially suitable for mobile tasks (e.g. route planning, pedestrian or car navigation, etc.). Research projects, for example, MUST (Almeida et al. 2002), SmartKom-Mobile (Wahlster 2002), MATCH (Johnston et al. 2002) and TeleMorph (Solon et al. 2004) use PDA devices with which the user can interact through a client-server architecture. In this setting the PDA (client) carries out the system input and output, whereas the server processes the input signals and generates the system response. Both client and server communicate through wireless connections. The response generation in this kind of scenarios implies a greater complexity than when an ordinary desk-top device is used, since current mobile devices are very limited in terms of memory, processing power, input modalities, wireless communication bandwidth, etc. For example, the TeleTuras system, implemented in the TeleMorph project, selects dynamically the output modalities, taking into account the available bandwidth when transforming the semantic representations provided by the dialogue manager into the information chunks to transmit through the diverse output modalities. If the available bandwidth is low, neither video nor animations are used in the mobile device, since otherwise it would negatively affect the interface quality, its effectiveness and the user acceptance. The input for the response generation module is a semantic representation based on the SMIL language (Synchronised Multimedia Integration Language).

In the case of the SmartKom system (Wahlster 2002), the input for the response generator is created using the M3L language, which is independent of the output modalities. This way, the output can be adapted to different scenarios depending on the user preferences (e.g. speech output is preferred by car drivers), features of the devices (e.g. display size) and language (English or German). To transform the M3L representation into the output according to these requirements, the system uses a set of XSLT patterns.

2.3 Output Interface

In the past 40 years many devices have been developed to stimulate user senses but only some of them have become popular for the general public. In the field of dialogue systems, these devices can be grouped into three categories depending on the user senses stimulated: *visual*, *acoustic* and *tactile/haptic*. The combination of information provided to the user through these devices allows the user to create a proper mental model of the performance of the system and the dialogue status, which facilitates the error correction and enhances the interaction. In this section we give a brief introduction to the technologies used to carry out the graphic generation, NLG, speech synthesis, sound generation and tactile/haptic generation.

2.3.1 Graphic Generation

Graphic generation is concerned with showing visual objects on the system screen, basically the system's animated agent and other graphic information about the task to be carried out (e.g. city maps showing the location of houses, restaurants and other facilities; tourist information with pictures and/or videos about points of interest; pictures of microwave ovens or other available products for sale, etc.). Additionally, the graphic generation is used by several systems to provide visual feedback about the key data obtained from the user, for example, the maximum price to pay and other requirements for the house to rent (e.g. number of bedrooms, swimming pool, terrace, garage, etc.).

2.3.1.1 Animated Agents

Many MMDSs include in the output interface a life-like character, usually called an *animated agent*, *conversational agent* or *embodied agent*, to achieve greater intelligibility and understanding of their generated spoken messages, especially when there are problems in

AdApt   August   Rea   Olga

Figure 2.9   Some animated agents (reproduced by permission of Gustafson et al. 2000, Centre for Speech Technology 1999, the AAAI Press and Beskow 1997, respectively)

the perception of speech due to noise or cross-talk (Cassell et al. 2000). In this section we present a brief introduction to these agents, which is completed in Section 3.2.4 where we discuss face and body animation standards and languages.

In the past two decades, a great variety of animated agents have been developed using several techniques (Figure 2.9). For example, Gustafson et al. (2000) used an approach based on 3 D parameterised and deformable models controlled by rules. These rules were used by a text-to-speech converter to analyse the input text and decide the values for the animation parameters, taking into account coarticulatory (e.g. lip rounding) and not coarticulatory (e.g. eyebrow raising) effects.

Several studies show the advantages and preferences of users for dialogue systems that include animated agents in their output interface, and others indicate that users are more likely to be collaborative if the animated agent has a human appearance (Koda and Maes 1996; Kiesler and Sproull 1997). Typically, the agent makes facial expressions, which are very useful to regulate the turn taking and help in carrying out other communicative functions, such as providing feedback about the internal state of the system (e.g. listening, thinking, etc.). Also, the agent makes the system output more intelligible, which is specially important in noisy environments and for users with hearing difficulties. Moreover, the agent gives the system an apparent personality, which may increase the sensation in the user of interacting with a real person and thus make him more cooperative.

For example, the animated agent of the Olga system (Beskow and McGlashan 1997) looks at the objects when the system references them by the spoken output, which increases the intelligibility of the spoken messages. It can make a limited set of gestures, such as pointing to specific areas of the screen, looking at the user, raising its eyebrows, smileing, etc. To select the output modalities (e.g. speech, graphics or speech and graphics) the system uses a set of rules that considers the output type. If the system understands the user sentence, the agent nods its head, or otherwise raises its eyebrows and asks for clarification. If the system does not obtain any data from the database query, the agent informs the user about that through a spoken message accompanied by a sad expression. The information obtained from the database is provided to the user through spoken and graphic messages, and detailed information about a particular product is shown graphically on the screen accompanied by a spoken summary.

The August system (Gustafson et al. 1999) also uses a set of rules to decide a variety of non-verbal behaviour, including gestures to show it is paying attention or thinking, as well as gestures to generate the six basic expressions (surprise, fear, disgust, anger, happiness and sadness) and transmit emotional information. Then rules take into account a variety of combinations of eyebrows and head movements to avoid them become repetitive. For example, a sentence can be accompanied by raising eyebrows followed by a slight head nod to emphasise an important word, or a slight head rise followed by an eyebrows movement to emphasise some syllables. Generally, the sentences generated by the system end with a slight head nod by the agent. Also, the agent generates a small set of gestures to show it is paying attention or thinking; for example, when the user presses the 'Push & Talk' button, the agent selects randomly one gesture of type 'paying attention'.

2.3.1.2 Other Graphic Information

In addition to the animated agent, MMDSs generally use other types of graphic information to be more reactive and informative for the users. This is typically the case of systems developed for the real estate domain, given that an apartment is a complex object for which it is suitable to provide information graphically (e.g. location in a city map, external appearance, etc.) and also orally (e.g. price, interior details, etc.). Several systems have been developed for this application domain, for example, Rea (Bickmore and Cassell 2004) and AdApt (Gustafson et al. 2002). The former shows on-screen images of the apartment the user is interested in, while the second uses graphic information to provide feedback about the data it has obtained from the user, which allows him confirm his input has been correctly understood by the system. It also shows graphically the constraints used to query the database (e.g. that the price must be below a specific amount) as well the locations of the apartments under discussion, which are marked on a map.

In another application domain, the Waxholm system (Carlson and Granström 1993) provides information about boat traffic in the Stockholm archipelago, using graphic information in the form of tables complemented by speech synthesis. As in the case of the AdApt system, it also provides graphic feedback about the information obtained from the parsing of the user utterance (e.g. time, departure port, etc.).

Graphic information is also typically used by systems developed to provide information about tourist places or city facilities, as it is the case for example of the TeleTuras (Almeida et al. 2002), MATCH (Johnston et al. 2002) or August (Gustafson et al. 1999) systems. The former uses a map to show a general view of the tourist places, as well as detailed maps with the points of interest in the centre. All the objects in the map that can be selected by the user are shown in a green square, whereas the objects selected are shown in a red square, accompanied by information about hotels and restaurants. Similarly, the MATCH system uses the graphic interface to show lists of restaurants as well as dynamic maps that show the localisation of specific points of interest in the streets. The August system shows 'thought balloons' on the screen to provide information not provided orally, for example, additional information or advice about what to ask the system. Also, it shows graphically information obtained from the database, e.g. restaurants and other city facilities.

Visual information is also used in other application domains, ranging from interaction with mobile robots to information about home devices. For example, the WITAS system (Lemon et al. 2001) shows in its GUI a map that displays the current operating environment of an autonomous and mobile robot, showing route plans, waypoints, etc. to a human operator. In another application domain, the Olga system (Beskow and McGlashan 1997), developed to provide information about microwave ovens, graphic information is used to provide detailed data about ovens while the system's animated agent provides summarised information using speech. The graphic information includes illustrations and tables that can be referenced by the agent.

Icons are a central part of today's graphical interfaces since they are very useful to remind users of basic functions of the applications and give them simple access to these functions. There are several reasons for using icons on the graphical output of MMDSs. First, they allow feedback to be provided in a non-linguistic way about what the dialogue system has understood from the user utterances, which does not increase user cognitive load. Second, users are good at automatically picking up the meaning of iconographic symbols presented to them repeatedly while they are doing other tasks, which facilitates the naturalness and efficiency of the interaction. Finally, icons can be associated with application constraints so that users can either manipulate such constraints graphically or multimodally by selecting them at the same time as speaking. For example, the Olga system uses icons to carry out specific actions (e.g. print data), whereas the MATCH system uses them to provide visual information to the user.

2.3.2 Natural Language Generation

Natural language generation (NLG) is a field of computational linguistics that deals with the creation of text responses from the internal representation of information handled by a computer program, which in our case is a dialogue system. NLG is not a trivial problem and there are many issues to address in order to allow the system to generate natural sentences. Among others, the system must select the words and expressions to use in a particular context. For example, the amount concept 'many, much' is expressed differently in expressions such as 'loud voice', 'work hard', etc. Another problem is concerned with correctly using the ellipsis and anaphora phenomena in sentences, since pronouns or substantives are used to take the context into account. Traditionally, the NLG is divided into two sequential phases, known as *deep* and *syntactic* (or *surface*) generation (Hovy 1993). The former can be divided into two phases: *selection* and *planning*. In the selection phase the system selects the information to provide to the user (typically obtained from a database query), while in the planning phase it organises the information into phrase-like structures to achieve clarity and avoid redundancy. An important issue of this phase is the lexical selection, i.e. choosing the most appropriate words for each sentence. For instance, a vehicle can be referred to in different ways, as for example, 'John's car', 'John's sport car', 'the car', 'the red car', etc. The lexical selection depends on the information previously provided by the system, the information available in the context, or specific stylistic considerations. The syntactic (or *surface*) generation takes the structures created by the deep generation and builds the sentences expressed in natural language, ensuring they are grammatically correct.

Several NLG systems can be found in the literature, characterised by different sophistication and power of expression (Baptist and Seneff 2000; Baptist 2000). Basically, they can be classified into four main groups: *canned, pattern-based, phrase-based* and *feature-based* systems. Canned systems are typically used in software products to generate error messages, and are very easy to build although they are very inflexible. Template-based systems represent the following level of sophistication. They are generally used when the same type of messages must be generated, but with slight differences. Each pattern contains several gaps that are filled with specific words (e.g. obtained from a database query) during the sentence generation. In the following level of sophistication are the phrase-based systems, which use generalised patters, i.e. patterns that contain other patterns. During the NLG each pattern is divided into other more specific patterns, which are again divided recursively until each sub-pattern is finally transformed into words. These systems can be very powerful and robust but are difficult to build for complex applications in which it is difficult to establish the valid relationships between patterns. If these relationships are not carefully tailored, wrongly formed sentences may be generated. Finally, the feature-based systems are among the most sophisticated systems for NLG. Each sentence is considered to be built from a set of simple features; for example, a sentence can be in different modes (e.g. affirmative, negative, imperative, interrogative, etc.) and different tenses (e.g. past, present, future, etc.). This way, the NLG is carried out by selecting the most appropriate features for each sentence. The main problems of this approach are in the feature selection during the generation and the maintenance of the relations between the features. For example, in the COMIC[4] project, XSLT patterns are used to transform the high-level representations generated by the dialogue manager into logical forms that are sent to the surface generator OpenCCG (White 2004). XSLT is used for planning many tasks, for example, structure and aggregation of contents, lexical selection and generation of multiple messages for the same internal representation when possible. OpenCCG is partially responsible for deciding the order and inflection of words, as well as for ensuring the correspondence between subject and verb (e.g. between *motifs* and *are*, *it* and *has*, etc.).

2.3.3 Speech Synthesis

Human fascination with creating machines able to generate a human-like voice is not merely the product of recent times. In fact, the first scientific attempts were carried out in the eighteenth century, using devices that replicated the form and sound-generation mechanism of the human vocal tract. One of them was a talking device developed by C. G. Kratzenstein in 1779 for the Imperial Academy of Saint Petersburg. This device generated vocal sounds as an operator blew air through a reed connected to a resonance chamber. Another attempt was a machine developed in 1791 by W. von Kempelen, which could generate the sounds of complete sentences. This device used blowers to make compressed air pass through a reed and excite resonance chambers. The device was controlled by a human operator. The operator fingers of one hand manipulated the shape of a resonance chamber to generate vocalic sounds, while the fingers of the other hand manipulated other resonance chambers to generate consonant sounds.

At the end of the nineteenth century, H. L. F. von Helmholtz and other researchers concluded that it was possible to generate voice-like sounds using electronic devices instead of mechanisms to replicate the human vocal tract. At the beginning of the twentieth century, J. Q. Stewart built a machine that used two electronic resonators excited by periodic pulses, and was able to generate different vocalic sounds by configuring appropriately the resonation frequencies. In 1930, H. Dudley, R. Reiz and S. Watkins developed a machine called *voder* that was exhibited at the 1939 World's Fair held in New York. This machine was the

---

[4] Conversational Multimodal Interaction with Computers (http://www.hcrc.ed.ac.uk/comic/)

first electric device able to generate speech and represents the basis for current speech synthesisers. However, synthesisers generate speech automatically, while the voder was manipulated by a human operator who used a kind of keyboard and a pedal to control the speech generation. In 1951, F. S. Cooper, A. M. Liberman and J. M. Borst used light and spectrograms (i.e. visual representations of the speech signal) to generate speech. They built a machine using a light source and a rotating wheel with 50 variable-density concentric circles to generate the different harmonics of the speech signal. This machine was able to transform into voice the information encoded in the spectrograms.

Current speech synthesis techniques can be classified into *voice coding*, *parametric*, *formant-based* and *rule-based*. In brief, the voice coding relies on sentences that are previously recorded, transformed into a bit sequence by an A/D converter, coded and stored in a ROM memory. The synthesis process is rather simple: the stored sentences are decoded, transformed into a analogic signal by a D/A converter and finally transformed into voice via a loudspeaker (Barnwell et al. 1996). Examples of voice coding techniques are PCM (Pulse Code Modulation), DM (Delta Modulation), DPCM (Differential Pulse Code Modulation), ADPCM (Adaptive Differential Pulse Code Modulation) and ADM (Adaptive Delta Modulation).

The parametric synthesis uses signal elements that can easily be modified using the so-called *source/filter* model, which models the excitation of the vocal tract and the articulation (Beskow 1998). This model is controlled by several hundreds of phonetic rules created considering the dynamics of the human vocal tract.

The formant-based synthesis is based on small units called *formants* (Puckette 1995). Each formant represents a group of harmonics of a periodic waveform produced by the vibration of the vocal cords. The vibration is carried out using electronic resonators tuned at different frequencies. Each phoneme is obtained by combining a determined set of formants. Speech synthesis is carried out by combining and changing properly the frequencies of the electronic resonators to obtain the corresponding phonemes.

The rule-based synthesis can make reference either to *concatenative* or *text-to-speech* (TTS) synthesis. Concatenative synthesis relies on previously stored voice units that are put together to obtain synthesised speech. A key question regarding this method is how to decide which units to use (e.g. sentences, words, syllables, semi-syllables, phonemes, diphonemes, etc.), which we discuss here briefly. The greater the unit size, the greater the number of units needed for the synthesis, and thus the memory requirements. Currently, using sentences as synthesis units allows the generation of messages of high intelligibility and naturalness. This unit is appropriate for applications that require a small number of sentences to be synthesised, but does not allow generation of an unlimited number of messages, nor messages that have not been previously recorded. The same applies to using words, which are useful when the application requires a limited vocabulary (a few hundreds of words as maximum) and new words are not added frequently.

Smaller synthesis units (e.g. syllables, semi-syllables, phonemes or diphonemes) provide greater flexibility and reduce the number of required units, since combining them makes it possible to obtain greater units and then synthesise an unlimited number of sentences. The semi-syllable is a fragment of speech that begins at the beginning of a syllable and ends at the middle of the syllable's vocal, or begins at the vocal's middle and ends at the syllable's end. Using this unit involves having pre-stored the most common prefixes and suffixes, which is important for the TTS synthesis that carries out an initial morphological decomposition. Another candidate synthesis unit is the phoneme; however, phonemes are not very appropriate since their acoustic manifestations do not exist as independent units, given that they are affected by the coarticulation of surrounding sounds. Therefore, their concatenation is more difficult and requires modelling the effect of the human coarticulation in the unit frontiers. A typically used unit is the so-called *diphoneme* or *diphone*, which is a small fragment of speech that contains the stationary part of a phoneme, the transition to the following phoneme, and the stationary part of this phoneme. The main advantage of using this unit is that it resolves the coarticulation problem, since the frontiers of the diphones contain only stationary states. Another common synthesis unit is the so-called *triphoneme* or *triphone*, which is also a context-dependent phoneme where the context is limited to the immediate left and right phonemes. The quality of the concatenative synthesis depends on several factors. One is the type of units chosen. They should allow as many coarticulation effects as possible and be easily connectable. Also, the number and length of the units should be kept as small as possible. However, since longer units decrease the concatenation problems, some trade-off is necessary.

A text-to-speech (TTS) synthesiser can be defined as a computer program that transforms into speech any input sentence in text format (Dutoit 1996). This feature makes this synthesiser very appropriate for dialogue systems, since it makes it possible to synthesise any sentence without needing to have pre-recorded words in advance. However, although very simple for human beings, the task to be performed by such a system is very complex due to a variety of reasons. One is the existence of abbreviations and other sequences of letters which cannot be transformed into speech directly (e.g. Mr., Mrs., Ms., etc.). Another reason is that the pronunciation of some words is not always the same as it may depend on several factors such as the position in the sentence (beginning vs. end), the type of sentence (declarative vs. interrogative), etc.

A conceptual TTS system contains two modules. The first carries out a *linguistic analysis* of the input text to transform it into a sequence of phonemes (or other basic units) accompanied by prosody marks that indicate how to pronounce them in terms of intonation, volume, speed, etc. The second module carries out a *digital signal processing* (DSP) to transform the output of the previous module into the output speech signal (i.e. the synthesised voice). The task carried out by the linguistic module can be implemented using six modules that work sequentially. The first is a *pre-processor* that identifies numbers, abbreviations (e.g. Mr., Mrs., Ms., etc.) and other particular letter-sequences and transforms them into pronounceable words. The second module is a *morphological analyser* that processes each word to find the parts-of-speech it contains on the basis of its spelling, and decomposes inflected, derived and compound words using grammars. For example, a word in plural (e.g. 'sandwiches') is decomposed into two parts-of-speech: 'sandwich' + 'es'. The third module is a *contextual analyser*, which considers the words in their context and reduces the list of their possible parts-of-speech to a very restricted number of highly likely hypothesis. The fourth module is a *syntactic-prosodic* parser, which examines the sentences and finds the text structure, i.e. the organisation into clause and phrase-like constituents that more closely relate to its expected prosodic realisation. Next, a *letter-to-sound* module automatically determines the phonetic transcription of the input text. Finally, a *prosody generator* includes the necessary prosodic features to create a segmentation of the input sentence into groups of syllables, which indicate the relationships between that groups.

2.3.4 Sound Generation

We are accustomed to receiving acoustically large amounts of data from the environment, either spoken or in the form of other sounds, which is very important to orientate and inform us about the state of the environment. Therefore, it is very important that MMDSs can communicate with users using this modality. The audio output in the form of speech is usually generated from the internal representation of the information handled by the system, as discussed in the previous section. But in addition to speech, a MMDS can generate other types of sound using a variety of methods. For example, some authors have used the so-called *auditory icons* (Gaver 1994), which are associations of everyday sounds to particular events of the system (e.g. a trashcan sound can be associated with the successful deletion of a file). Another possible method is to use the application data to control the sound generation (i.e. data can be mapped to pitch, brightness, loudness, etc.). For example, changes in the bond market data can be associated with changes of brightness of the generated sounds (Kramer 1994).

2.3.5 Tactile/Haptic Generation

Current MMDSs typically make use of visual and acoustic modalities to provide information to the user without considering tactile/haptic output. However, this output modality could be used to increase the information obtained from the system in particular applications (e.g. virtual reality or interaction with robots). According to Shimoga (1993), available tactile/haptic devices can be classified into four categories: *pneumatic, vibrotactile, electrotactile* and *functional-neuromuscular*. Pneumatic stimulation includes devices that use air jets, air pockets or air rings to stimulate the haptic sense of the user. The main drawback of these devices is that they cause muscular fatigue and have low bandwidth. Vibrotactile stimulation includes devices that produce vibrations using blunt pins, voice coils or piezoelectric crystals. These devices seem to be the most suitable to stimulate the somatic senses of the user since they can be very small, lightweight and achieve high bandwidth. Electrotactile stimulation includes devices that use small electrodes attached to the user's fingers to provide electrical pulses. Finally, functional-neuromuscular devices directly stimulate the neuromuscular system of the user, which makes them inappropriate for standard users.

2.4 Summary

In this chapter we have introduced the main technologies employed to set up dialogue systems, classified according to the module structure of the conceptual MMDS shown in Figure 1.1. The chapter is divided into three parts, corresponding respectively to the input interface, multimodal processing and output interface. The first part discussed the ASR technology, addressing a brief history, classifications of ASR systems, problems and types of error. It also discussed the stochastic approach as it is probably the most used at the moment, addressing acoustic and language modelling. Next, the chapter outlined the NLP, providing a brief description of the analysis types, main problems and available analysis grammars. Then, the chapter addressed the face localisation and tracking, discussing approaches based on knowledge, invariant features, pattern matching and colour. It focused on gaze tracking, discussing applications, benefits for users and implementation techniques, focusing on some non-intrusive tracking methods. Next, the chapter addressed the lip-reading recognition, commenting on its effect on speech perception (cf. McGurk and ventriloquist effects), and some lip-reading techniques. Gesture recognition considered the gestures made in the 2D and 3D space. Recognition methods were discussed for intrusive and non-intrusive devices, and gesture classifications were commented on. Then, the chapter focused on handwriting recognition, discussing its advantages and similarities with ASR (e.g. user dependence/independence, existence of different possible recognition units, vocabulary size, etc.). Available devices and recognition methods based on structural analysis, statistics and trajectories were discussed.

The second part of the chapter dealt with the technologies employed to set up the modules in the multimodal processing component of Figure 1.1. It studied the multimodal data fusion module, describing the features of fusion at the signal and semantic levels, and discussing methods available for setting up the fusion process based on ANNs, HMMs, FFTs, etc. Then, the chapter focused on the multimodal data storage module, briefly discussing several formalisms typically used to represent multimodal data (melting pots, frames, XML-languages and TFSs). The next module addressed the dialogue manager. A very brief discussion was provided since Chapter 5 is specifically concerned with dialogue modelling and management. The chapter addressed briefly the task module, discussing benefits and how several researchers have used tasks-independency to set up dialogue systems. It also addressed briefly the database module, commenting types (local vs. remote) and some possible contents (application data, user data and emotional data). This part of the chapter finished by discussing the response generation module, which carries out the *fission* operation that decides the output modalities to use to present the information to the user appropriately. The importance of the presentation mode was discussed, commenting on the limitations imposed by mobile devices (e.g. PDAs), and describing how the fission is carried out in several MMDSs set up using these devices (MUST, SmartKom mobile, MATCH and TeleTuras).

The final part of the chapter dealt with the technologies employed to set up the modules in the output interface of the conceptual MMDS studied. It began with a description of the graphic generation focusing on the animated agents, discussed the main advantages of these agents and, as a sample, commented on the features and performance of the agents employed in the Olga and August systems. Then, the chapter addressed other types of graphic information (e.g. city maps, pictures of apartments, etc.), discussing, as a sample, how this information is handled by the AdApt and Rea systems for the real estate domain, the Waxholm system for boat traffic information, and the MATCH, August and TeleTuras systems for tourist information. Finally, it discussed the benefits of using icons in the GUI, describing as a sample how they are used in the Olga and MATCH systems. Next, the chapter dealt with the NGL module, describing the main problems (e.g. lexical selection, anaphora, ellipsis, etc.), the traditional phases to generate sentences (deep and surface generation), and finally the diverse systems available to carry out the NLG (canned, pattern-based, phrase-based and feature-based). The chapter then focused on speech synthesis technology, with a brief description of the historic antecedents (acoustic, electro-acoustic and electronic devices) and discussing the current speech synthesis techniques (voice coding, parametric, formant-based and rule-based). Next, the chapter discussed sound generation techniques additional to speech, such as the so-called *auditory icons*, which are associations of everyday sounds with particular events of the system. To conclude, the chapter briefly introduced the technology concerned with the generation of tactile/haptic output, which can be carried out by means of pneumatic, vibrotactile, electrotactile or functional-neuromuscular devices, and may be very useful in virtual reality and robot-based applications.

2.5 Further Reading

Input Interface

- Speech processing and ASR: Chapters 5–13 of Huang et al. (2001).
- General information about NLP: Allen (1995).
- NLP information focused on dialogue systems: Chapter 17 of Huang et al. (2001) and Chapter 4 of McTear (2004).
- Face localisation: Chapter 5 of Gong et al. (2000).
- Gaze tracking: Duchowski (2003) and Chapter 8 of Gong et al. (2000).
- Technology of input system modules: Section 2.8 of Gibbon et al. (2000).

Multimodal Processing

- Multimodal data fusion: Sections 2.4 and 2.6 of Gibbon et al. (2000).
- Dialogue management: Part 1 of Taylor et al. (2000) and Chapter 5 of McTear (2004).

Output Interface

- NLG: Chapter 4 of McTear (2004).
- TTS and speech synthesis: Chapters 14–16 of Huang et al. (2001).
- Technology of output system modules: Sections 2.7 and 2.8 of Gibbon et al. (2000).

3

Multimodal Dialogue Systems

3.1 Benefits of Multimodal Interaction

The benefits of multimodal interaction can be considered in terms of system input, processing and system output. This section begins discussing the advantages in terms of system input, which are concerned with the existence of several interaction modalities from which users can choose to fit their preferences and/or environmental conditions. Then it presents the advantages of multimodal processing, focusing on the compensation of the weaknesses of the individual modalities, and on the resolution of multimodal references. Finally, the section outlines the benefits in terms of system output, concerned with the use of several modalities to provide the information requested by the user, as well as information about the system internal state. The section ends by discussing the supposed advantages of using animated agents in the output interface.

3.1.1 In Terms of System Input

One of the main features of MMDSs is the use of several communication modalities to enhance the interaction with the user. Thus, the user is not restricted to just a microphone or a telephone to interact with the system but can use additional input devices (e.g. keyboard, mouse, artificial vision camera, touch-sensitive screen, etc.), which makes MMDSs more flexible, powerful and effective than SDSs. For example, using a speech recogniser and a touch-sensitive screen, the user can move an object on the screen to a different location just by saying 'Put that there' while pointing to the object and to the new location. Moreover, deictic references to objects on the screen may be easier by pointing than by using speech, while the execution of commands may be easier by pronouncing them orally rather than choosing them from a long list on the screen using a pointing device.

MMDSs give users the chance to decide when and how to use a particular input modality in order to achieve an efficient interaction. Thus, users may avoid a modality that they consider might be error-prone (for example, they may avoid pronouncing a foreign city name and may prefer to point it out with a pen). Additionally, if a recognition error happens

Spoken, Multilingual and Multimodal Dialogue Systems: Development and Assessment Ramón López-Cózar Delgado and Masahiro Araki © 2005 John Wiley & Sons, Ltd

using an input modality, they may use a different modality to provide the data. Oviatt (1996) indicated that there is a tendency in users to switch from an input modality to another if the interaction with the system becomes problematic. In simulated experiments in which users were subjected to errors that required them to repeat their input up to six times, users changed from speech input to graphic input after they had rephrased their input several times, i.e. they used the modality change as a recover strategy after having been subjected to repeated errors.

Multimodal interaction gives the user the possibility to replace complex verbal expressions with non-verbal expressions, thus reducing lack of fluency and mistakes. Considering that a user may choose ways of expression that may not be the most suitable for the system, several authors have tried to identify the factors that determine the user selection of modalities. Several studies indicate that the system expressions influence the user expressions due to the tendency of human beings to coordinate their own way of expressing with that of the other dialogue partner. For example, Zoltan-Ford (1991) proved that the length of the user sentences is influenced by that of the system (a phenomenon called *verbal shaping*) whereas Brennan (1996) proved that the user vocabulary is also influenced by the system vocabulary (a phenomenon called *lexical shaping*).

If a task can be carried out using different modalities, users may differ in their preferences for one or another. Moreover, apart from preference, users may have different abilities to interact with the system, for example, handicapped users may be unable to use a keyboard or a mouse. Thus, they will be more comfortable using a system that allows them to make the choice verbally.

The availability of several input modalities represents a great advantage since speech input is not always the best input interaction modality. MMDSs allow the interaction to be adapted to the environmental conditions in terms of noise, light, privacy concerns, etc., allowing the users to select one modality or another depending on those conditions. For example, speech interaction may not very appropriate in noisy environments or when there are other people talking nearby, so that it may be preferable to use handwriting instead. Oviatt and Olsen (1994) studied the selection of these two modalities (speech and handwriting), finding that there were several factors determining the selection of one or another, as for example:

- *Content of information.* Digits were more likely to be written than text, and the same applied to proper names in comparison to other types of word.
- *Presentation format.* Data requested to the user via forms were more likely to be written than spoken.
- *Contrastive functionality.* Some 75% of the pen/voice patterns were a consequence of the user preferring to change the input modality to indicate a context change or a communicative function, such as providing data/correcting data, data/command, digit/text, etc.

Woltjer et al. (2003) also studied the user selection of modalities, considering an in-car MMDS for e-mail access. Taking into account that users may have problems in combining driving with the system interaction, the study compared the differences between three interaction types: oral, manual/visual, and modality chosen by the user. The results showed that the highest users' subjective load was achieved for the manual/visual interaction, and the lowest when they were able to choose the interaction modality.

3.1.2 In Terms of System Processing

Several authors indicate that users not only prefer MMDSs to SDSs, but also that the multimodal interaction reduces the error rates and the time required to carry out the tasks, increasing effectiveness and user satisfaction (Granström et al. 1999, 2002; Terken and Te Riele 2001). For example, Oviatt (1996) showed experiments with interactive maps using three types of interaction modalities: speech only, pen only, and multimodal (speech and pen). In comparison with the speech only interaction, the multimodal interaction reduced by 10% the time needed to complete the tasks, by 35% the number of disfluencies, and by 36% the number of task errors.

Multimodality increases the robustness of dialogue systems by integrating synergistically the available input modalities in order to disambiguate and combine the information chunks obtained from the user input, which compensates (to some extent) for the limitations of the technologies employed. For example, speech can compensate for some limitations of graphic interfaces by making it possible to reference objects that are not on the screen at a given moment of the dialogue. Also, graphic interfaces can compensate for some limitations of speech by making visible the results of the actions on the objects, and showing what objects and/or actions are really important for a task at a given moment of the dialogue. Additionally, the errors made by a device can be compensated for with the information provided by other devices, which alleviates some of the problems of SDSs. For example, let us consider a system that combines speech and 2D gestures, and suppose the user says 'What is written here?' while he encircles an area on the screen (Corradini et al. 2003). Let us suppose the speech recogniser generates the hypotheses 'What is grey here?' and 'What does it say here?' together with the correct hypothesis. Suppose also the gesture recogniser generates the hypotheses 'the user wrote the letter Q' and 'the user drew a circle'. The multimodal data fusion module must choose the most likely interpretation of the user intention from these hypotheses, and send it to the dialogue manager. The choice can discard the inconsistent information by binding the semantic attributes of both modalities; for example, only the second gesture hypothesis can be accepted if the spoken command can only be combined with an 'area selection' gesture.

Using the same strategy, Oviatt (2000) showed that the recognition of the user intention can be increased by combining the N-best outputs of the speech and gesture recognisers in the QuickSet system. The author shows an example in which the user says 'zoom out' and draws a checkmark. The utterance is ranked fourth in the speech N-best list, whereas the checkmark is correctly recognised and appears first in the gesture N-best list. The combination of both information chunks achieves the correct user intention (zoom out) since the three hypotheses with the higher score in the speech N-best list only can be integrated with a circle or interrogation gestures, which are not on the gesture N-best list. Consequently, they are discarded during the fusion process.

Johnston et al. (1997) also used the information provided by the gesture recogniser, but to compensate for the spurious errors of the speech recogniser when the user was not speaking but the speech recogniser provided words (*false alarms*, discussed in Section 2.1.1). The compensation method was simple but effective. If the recognised words generated a semantic representation that required a 'location' data to be complete, and this bit of data was not provided by a gesture simultaneously or immediately, then the speech information was discarded.

These examples clearly show the benefits of cooperation between the input modalities. According to Martin et al. (1998), the cooperation can be of six types:

1. *Complementarity*. Different information chunks are transmitted through more than one modality. For example, in terms of system input, the user may say 'Put that there' while pointing to the object and then to the location. In terms of system output, some information may be provided to the user by speech (e.g. position, shape, size of objects), while other information is provided by deictic gestures of the system's animated agent pointing at pictures or icons on the screen.
2. *Redundancy*. The same information chunk is transmitted through more than one modality. For example, in terms of system input, the user may say 'I want the second option on the right' while he points at that option with the mouse. In terms of system output, the same information is provided by speech and deictic gestures of the animated agent when possible.
3. *Equivalency*. An information chunk can be transmitted through more than one modality. For example, in the system input a menu option can be selected either orally or using the mouse, whereas in the output specific information can be given either by speech (e.g. 'The object in the centre') or by a deictic gesture of the agent pointing at the object.
4. *Specialisation*. A specific information chunk is always transmitted through a determined modality (e.g. speech). In this case, the gestures made by the animated agent are only rhythmical.
5. *Concurrency*. Independent information chunks are transmitted using interaction modalities that overlap in time. For example, the user speaks on the phone while editing a document.
6. *Transfer*. An information chunk produced by a modality is transformed into another modality. For example, the user describes images using speech.

As commented above, multimodal interaction offers diverse advantages that enhance the user–system communication. However, using several information sources can also cause ambiguity, uncertainty or contradictions. An example of ambiguity occurs when using a pen on a touch-sensitive screen, the gesture made can be interpreted by several recognisers of the system, e.g. the gesture and the handwriting recognisers, which will create their own recognition hypotheses. Moreover, even if a gesture is interpreted by only one recogniser, it can have several meanings, in which case using another modality (e.g. speech) can be useful to make the disambiguation. For instance, an arrow made on the screen can be interpreted differently if the user says 'scroll map' or 'show this photo'. An example of contradiction occurs e.g. when interacting with a system developed for pedestrian navigation, set up in a PDA, the user says 'scroll map to the west' while he draws an arrow on the screen pointing to the east.

The cooperation between input modalities can also be used to resolve referring expressions. For example, let us suppose the pedestrian navigation system noted above can correctly interpret the sentence 'Show me the photo of the hotel on Main Street'. Thus, the system must find the hotel the user is referring, which can be done by combining the information provided by speech and gestures. For example, the gesture recogniser can indicate that the user has pointed to an object or has made a determined gesture on the screen while uttering the sentence, which may determine the referenced hotel. In absence of a gesture, the system's multimodal data storage module can provide information about the dialogue history, i.e. the objects that have been mentioned recently by the user. If the dialogue history contains the name of a hotel located on Main Street, this hotel may be considered the referenced one. Otherwise, the system can check its hotel database to know how many hotels are located at the specified address. If there is only one, the reference can be resolved, whereas if there are more, then the system should ask the user to clarify the reference.

3.1.3 In Terms of System Output

The multimodal interaction allows several communication channels to be used to provide information to the user (e.g. speech, text, graphs, images, etc.), stimulating several of his senses simultaneously (Oviatt 1996; Gustafson et al. 2000). This feature is useful for several kinds of application such as train/airplane information systems, which typically provide travel dates, timetables, etc. An MMDS can show graphically the key data to provide for the user (e.g. departure and arrival times, train connections, etc.), so that he can take notes and really appreciate the system. Also, the system can provide visual indications about its internal state (e.g. accessing the database, confused when processing the user input, etc.), which is useful to provide feedback to the user about the system performance.

Several authors have discussed the advantages of using an animated agent to increase the intelligibility of the system output. For example, House et al. (2001) showed that using such an agent in a noisy environment increased the global intelligibility of VCV stimuli by 17% (from 30% when only the speech synthesiser output was used to 47% when the synthesiser output was accompanied by the animated agent). However, there are also studies indicating that those advantages are questionable. Various works have been carried out in order to study objectively and subjectively the validity of the supposed advantages, and some of them have shown that using an animated agent does not necessarily lead to interaction enhancement. For example, Walker et al. (1994) and Takeuchi et al. (1995) indicated that these agents require a greater effort and attention on the part of the users to interpret their behaviour, thus using them is not necessarily the best option. Koda and Maes (1996) indicated that animated agents help users engage in the tasks and are appropriate for entertainment applications. They also indicated that the user opinions about an animated agent were different if it was considered an isolated occurrence or in a given task, perceiving the system 'intelligence' by its behaviour in the task and not by its appearance. Van Mulken et al. (1998) studied the effects of animated agents, taking into account subjective and objective measures (comprehension and recall) using the PPP Persona agent (André et al. 1996) both in technical and non-technical information tasks. The subjective evaluation results indicated that the so-called *persona effect*[1] was observed in the technical information domain but not in the other one, without affecting the objective evaluation measures. Koda and Maes (1996) indicated that there are users in favour and against personification, and that the differences in terms of facial features, gender, expressivity, etc. of such agents cause different effects on both groups. For example, they observed that users in favour attributed more intelligence to an expressive face, while those against considered a stoic face was more intelligent. Therefore, given the dichotomy of results in favour and against, it seems reasonable to think that MMDSs should be flexible to support a diversity of user preferences and task requirements. From the studies carried out, it also follows that animated agents should be used in application domains in which it is important to transmit to the users the agent behaviour so that they feel comfortable using the system, as well as in application domains in which it is vital to attract the user attention.

---

[1] This effect refers to the general supposed advantage of an interface that includes an animated agent (in comparison with the same interface without the agent).

3.2 Development of Multimodal Dialogue Systems

MMDSs of different complexity have been developed for more than three decades. Initial systems, for example, Scholar (Carbonell 1970) or NL (Brown et al. 1979) combined written natural language with simple gestures made on the screen. Subsequent systems combine speech and gesture input, for example, Put-That-There (Bolt 1980), Talk and Draw (Salisbury et al. 1990), Cubricon (Neal and Shapiro 1991), Matis (Nigay et al. 1993), VoicePaint (Salber et al. 1993), Vecig (Sparrel 1993), AdApt (Gustafson et al. 2000), etc. There are systems that are multimodal only in terms of the output being the input unimodal (spoken), for example, Olga (Beskow and McGlashan 1997), Waxholm (Carlson and Granström 1993), and August (Gustafson et al. 1999).

Even for restricted domains, the setting up of dialogue systems for real-world applications is a very complex problem that is typically divided into sub-problems associated with system modules that carry out specific tasks. These modules are generally grouped into input and output interfaces, interconnected by a module (dialogue manager) that communicates both interfaces and implements the 'intelligence' of the system (as shown in Figure 1.1). According to Martin et al. (1998), some usability aspects to take into account in the development of MMDSs are the following:

- *Efficiency.* This property is mainly concerned with the time required to carry out the tasks satisfactorily. Generally, this property measures aspects such as the time required to either make a selection, identify the user intention or complete tasks.
- *Error prevention.* As in the case of SDSs, MMDSs must employ strategies to avoid interaction errors. Typically, confirmations are used for this purpose but at the cost of lengthening the dialogues, thus reducing efficiency as more time is needed to complete the tasks (due the additional confirmation turns). Thus, the system design must search for a trade-off between accuracy and efficiency (e.g. trying to set up an optimal confirmation strategy).
- *Information about the system state.* As in the case of SDSs, feedback information is very important for MMDSs to alleviate drawbacks or limitations of some modules. In the case of MMDSs, feedback can be provided to the user multimodally, typically by means of graphics, sounds and speech.
- *Ergonomic issues.* Ergonomics is a very important factor in dialogue systems, and the system design must take care of possible user fatigue caused by the interaction with these systems.

3.2.1 Development Techniques

3.2.1.1 WOz

The WOz (Wizard of Oz) technique is a simulation method in which a person called *wizard* simulates the behaviour of the system to be developed (or some parts of it), and the user interacts with the simulated system believing that he is actually talking to a dialogue system (Gould et al. 1982). Then, when he invokes a function that is not implemented in the system, the wizard simulates the effect of such action. This technique is very widely used due to several reasons. One is that implementing the complete dialogue system may be impossible or too costly. In this case, the WOz technique can be used to evaluate the usefulness and need for the system, before its setting up is started. Another reason is that the system designers may need to collect information about the application domain for which the system is developed, which can be used to observe the needs and behaviour of different users, the tasks to be carried out by the system, and the system features to develop. A third reason for using the WOz technique is to refine and/or check the design decisions made, which is very important to identify potential weak points before setting up the system in the real world.

Usually, the setting up of dialogue systems takes into account human dialogues in the domain for which the systems were designed, but since users behave differently when they speak to computers, the development should not rely exclusively on human-to-human conversations, but should also take into account knowledge about the specific characteristics of the human-to-computer dialogue. There is a variety of reasons for that. On the one hand, it has been observed that when a person interacts with another in a real dialogue, he tends to adapt his speech to fit the characteristics of the speech of his dialogue partner. For example, the language used by an adult to communicate with a child varies notably in comparison to the one used to communicate with another adult. Thus, it is reasonable to expect that a person will speak differently to a human operator than to a dialogue system. Also, the language is affected by a variety of interpersonal factors. For example, Lakoff (1973) indicated that in human dialogues the so-called *indirect speech acts* are common, produced as a result of courtesy rules. However, users of a dialogue system may not feel the need to be polite to the system. Given the differences between human-to-human and human-to-computer dialogues, it is clear that the design of these systems cannot rely exclusively on human dialogues, which may differ notably in terms of style, vocabulary and interaction complexity. Thus, it is very important to use empirical data extracted from human-to-computer dialogues, which can be obtained using the WOz technique.

In order to make this technique successful, it is very important that the wizard behaves appropriately. Although apparently simple, his behaviour requires a lot of knowledge in order to be consistent in terms of content, style and rhythm. Thus, the wizard must be properly trained to carry out well-defined tasks. However, for some applications it can be very difficult to have detailed guidelines for all the possible situations that can arise during the dialogue, in which case the wizard should use his good judgement to give an appropriate response. It is important that the wizard practises with the tool to use in the experiments, to be trained in using it before interacting with the users. Such a tool usually includes a set of predefined answers and menus containing parts of pre-recorded responses, which ensure that he can generate the 'system' responses quickly.

Several tools have been developed to set up the WOz technique, such as ARNE, developed by Dahlbäck et al. (1993), which incorporates an editor to create response patterns including canned text and gaps to fill with data obtained from database queries, another editor to define the database queries, tools for presenting graphic information, and a record of dialogue turns accompanied by timestamps. A window shows the wizard–user dialogue and a pop-up menu allows the wizard to select standard prompts. Another window allows the wizard to make database queries and visualise the results, allowing him to select the pattern gaps that must be instantiated with the data obtained from the query. The tool includes another window that allows the wizard to control the dialogue by selecting the current dialogue state as well as the next possible states.

The WOz technique has been used to develop several SDSs, for example, Schisma (Van de Burgt et al. 1996), Basurde (Vilar et al. 2003), Jupiter (Glass et al. 2000) and other systems for providing shop information while driving (Murao et al. 2003), operating a TV set to receive broadcasts (Goto et al. 2004), etc. However, although this technique is very useful and is typically used in the design of dialogue systems, it also has some drawbacks. One is that it is really difficult to simulate the behaviour of a dialogue system, especially the speech recogniser. Thus, the wizard behaviour must be carefully planned to avoid designing dialogue strategies that are not robust enough. Another problem is that the WOz experiments can take large amounts of effort and time, mostly due to the recruitment of the test users, the set-up of the experiments and analysis of the data obtained. Providing users with information about the task to carry out in the experiments is also a delicate issue since it can bias the way they will interact (e.g. the sentence types they will utter), so that in some scenarios this information is provided graphically to avoid influencing the user way of expression, vocabulary, etc. Also, if too much information is given to the users in the explanation of the WOz experiments, they may try to replicate the explanations or what they have seen in examples shown, instead of interacting in their own way.

3.2.1.2 WOz in Multimodal Systems

The WOz technique has also been used to develop MMDSs, for example, Waxholm (Carlson and Granström 1993), Jeanie (Vo et al. 1996), AdApt (Gustafson et al. 2000), SmartKom (Steininger et al. 2001), Mask (Life et al. 1996), etc. Applying the technique to these systems requires simulating more complex functions than in the case of SDSs, for example, the synergistic combination of modalities (Salber and Coutaz 1993). The diverse interaction modalities can cooperate in different ways (as discussed in Section 3.1.2) and can be processed independently or combined by the fusion process (studied in Section 3.2.2). Therefore, the workload in a multimodal interaction can be excessive for just one wizard, leading to inconsistent behaviour in terms of content, form or response time. For this type of scenario, a multiwizard configuration that allows a variable number of wizards and a variety of input and output devices is the proper solution. Also, since the organisation of the wizards requires a lot of testing and experiments, the communication protocol among the wizards must be flexible.

Salber and Coutaz (1993) proposed a multiwizard configuration in which the workload was divided among several wizards. For example, one wizard specialised in the input/output whereas another specialised in the dialogue management. The balanced distribution of the workloads may facilitate a consistent behaviour without negatively affecting the response time. However, given that it may be difficult to estimate the workload in advance (since it notably depends on the user behaviour), some wizards can dynamically change their roles. The authors proposed that a wizard could behave as a supervisor, his task being to observe the behaviour of the other wizards, control the session, and take the appropriate decisions in case of system malfunction. The authors developed a multiwizard platform called Neimo to facilitate the observation and evaluation of multimodal applications. The platform stores a great diversity of data regarding user behaviour obtained either automatically by the system or by the wizards. The platform was designed to carry out experiments using several modalities, facilitating the creation of log files to store data at different levels of abstraction (from keystroke/event level up to commands and domain concept levels). An essential requirement in the design of the tool was flexibility to allow any number of wizards and modalities, as well as any kind of application. Moreover, for a given application and a determined set of modalities, the tool experimented with information fusion/fission and time constraints.

Figure 3.1 shows the wizard interface used to develop the AdApt system (Gustafson et al. 2000). The user input was sent to this interface where the wizard decided the system response using a set of pop-up menus of patterns, in which he included the information about apartments obtained from database queries. In addition to the spoken input, the wizard had a window showing the user actions on a map (e.g. mouse movements, selections, etc.).

The interface included an animated agent, an interactive map showing the names of the main streets, neighbourhoods, parks, etc. The wizard response was very quick (usually 1 or 2 seconds) giving the impression that the users were really interacting with a real system. The animated agent showed a 'listening' facial expression when the user was speaking and it showed a 'thinking' expression when silence was detected. Within this reactive behaviour of the agent, the wizard created the system response using the pattern menus while the agent was showing the 'thinking' expression.

3.2.1.3 System-in-the-loop

On some occasions it may be convenient to build the system using a prototype that is being refined cyclically, in which case the dialogue structure is enhanced progressively by the user interaction. This technique has been used to develop several systems, such as Schisma, to provide information about theatre performances (Van de Burgt et al. 1996); Jupiter, to provide weather information (Zue et al. 2000; Glass et al. 2000); MATCH, to provide restaurant and subway information (Johnston et al. 2002), etc. The technique has several advantages. One is that it avoids the need to dedicate time and effort to implement the WOz experiments since the users interact with the real system, not with a simulation of it. Another reason is that it avoids the possible mistakes the wizard can make when interacting with the users. A third reason is that this approach collects speech material, tests and refines the

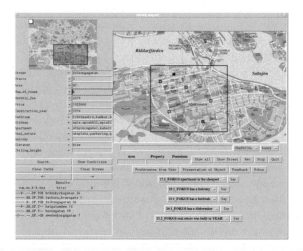

Figure 3.1   The wizard interface used for the development of the AdApt system (reproduced by permission of Gustafson et al. 2000)

system at the same time, and using the same signals, e.g. background noise, signal-to-noise ratio (SNR), etc., that the final system will process. Thus, there is a better match between training and testing conditions since all the components of the final system are used from the start, which facilitates finding their possible weaknesses before the final system is built. The disadvantage of this technique is that the initial prototype system must work well enough to be used for interacting with the users.

3.2.2 Data Fusion

As discussed in Section 3.1.2, MMDSs must process information chunks provided via several modalities. These chunks cooperate to transmit the user intention. To obtain such an intention, the chunks must be combined (*fusioned*). The multimodal data fusion module (Figure 1.1) is responsible for this task. Section 2.2.1 discussed the classical methods to carry out the fusion, either at the signal or semantic levels. Some alternatives to these classical methods have been proposed. For example, another approach is based on hybrid symbolic/statistical fusion methods, which attempt to improve multimodal recognition and disambiguation rates by using corpus-based statistics to weight the contributions from various input streams. To do so, Wu et al. (1999, 2002) proposed a fusion method called MTC (Members, Teams, Committee), where the 'members' represent the individual recognisers of the MMDS, the 'teams' train the conditional probabilities weights, and the 'committee' extracts a decision from the output of the 'teams'. This approach was used by Kaiser and Cohen (2002) in the QuickSet system to assign weights to the speech and gesture constituents of 17 general types of multimodal commands, taken from a multimodal corpus obtained for a previous study. From a test corpus containing 759 multimodal inputs, processed using the N-best recognition technique, the error rate in predicting the general type of the 1-best recognition hypothesis was 10.41%. However, using the MTC method to weight the contributions of both modalities (modelling in consequence their dependencies) reduced the error to 4.74% (a relative error reduction of 55.6%).

3.2.2.1 Fusion of acoustic and visual information

ASR is a difficult task in noisy environments. In order to increase recognition rates, it is possible to use recognition techniques that are robust against some types of noise, and/or use additional information that is not affected by noise. In this category, speech-reading (or in other words audio-visual ASR, AVASR) aims to use information about the user face during speaking. The advantage is that the visual modality is not affected by the acoustic noise and provides complementary information to the acoustic one. Speech recognisers of SDSs use diverse types of information to process the user sentences, mainly acoustic, grammatical and prosodic; however, optical information regarding the speaker face is not used. In noisy environments, human beings use a combination of both types of information, enhancing the SNR by a gain of 10 to 12 dB (Brooke 1990). The lip movements provide about two-thirds of the information, but there are other parts of the face which are very important to understand speech by the vision modality, for example, facial expressions and position of head, eyebrows, eyes, ears, teeth, tongue, cheeks, etc. Thus, the combination of all this visual information with the speech signal is very important to enhance the performance and robustness of MMDSs.

AVASR has become an attractive research field for the problem of speech recognition in noisy environments. Several approaches have been presented so far that can be divided into two groups: *feature fusion* and *decision fusion*, although there are also hybrid approaches that combine characteristics of both. In the feature fusion approach, the observation vectors are obtained by the concatenation of the audio[2] and visual[3] observation vectors. The fusion methods of this approach are based on training a single classifier (i.e. of the same form as the audio- and visual-only classifiers) on the concatenated vector of audio and visual features, or on any appropriate transformation of it. The drawback of this approach is that it cannot model the natural asynchrony between the audio and video features. The *direct identification* architecture presented by Rogozan and Deléglise (1998) is an example of method of this fusion category. It integrates the audio and video information in just one stage concatenating the acoustic and visual observations, as shown in Figure 3.2(a). The visual features are considered additional to the acoustic features without taking into account the specific nature of vision. Thus, it only uses one recognition unit (the phoneme) that does not take into account any difference when modelling the acoustic and visual information.

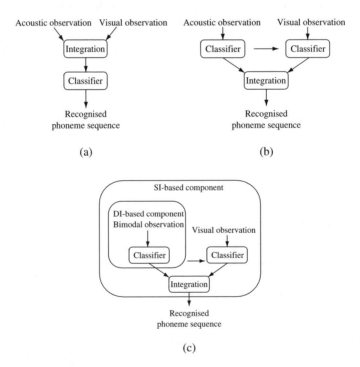

Figure 3.2   Three different architectures for audiovisual speech recognition (reproduced by permission of Elsevier)

---

[2] Audio vectors contain data obtained from the speech signal, e.g. linear predictive coefficients, filter bank coefficients, etc.

[3] Visual vectors contain data about the oral cavity, e.g. area, perimeter, height and wide, jaw opening, etc.

Therefore, this architecture cannot model the anticipation/retention phenomena,[4] given that the acoustic and visual data force the same borders between successive phonemes.

In the decision fusion approach, the audio and video sequences are modelled independently and the synchrony of both types of feature is enforced only at the model boundary. The fusion methods in this approach use two single modalities (audio- and visual-only) classifier outputs to recognise audiovisual speech. For example, Liu et al. (2002) proposed a method to integrate audio and visual observation sequences using a coupled Hidden Markov Model (CHMM) that models the audiovisual state asynchrony, and at the same time preserves the natural audiovisual dependencies over time. The *separate identification* architecture presented by Rogozan and Deléglise (1998) is an example of this approach. It allows a little asynchrony in the processing of acoustic and visual data since both types of data are combined at a higher phonetic level. The observations from both sources are processed in two separate but intercommunicated components, as shown in Figure 3.2(b). Rogozan and Deléglise also presented a fusion approach, called *hybrid identification* architecture, which replaces the acoustic component of the separate identification architecture by a component based on the direct identification architecture, Figure 3.2(c). In this architecture the integration takes the maximum advantage of the complementary nature of both modalities.

Other AVASR methods have been proposed. For example, Girin et al. (2001) proposed a method to enhance speech recognition using geometrical lip parameters based on a spectral estimation and a Wiener filter. The video parameters were used to estimate a filter that was applied to the noisy audio signals. Berthommier (2003) proposed a method to enhance audiovisual speech based on the association between speech envelope and video features. In this method, the idea is to observe the relationship existing between mouth, face movements and formant trajectories represented by line spectral pairs and an additional RMS (Root-Mean-Square) energy parameter. Daubias and Deléglise (2002) extracted visual information from the images of a speaker using shape and appearance models, which allow the adaptation to the task or environmental conditions. Other authors represent the oral cavity movements using derivatives, surface learning, deformable templates, or optical flow techniques.

A number of approaches have been proposed to implement these methods, such as histograms (Nock et al. 2002), multivalued Gaussians (Nock et al. 2002), ANNs (Meier et al. 2000), or HMMs (Nock et al. 2002). In all these approaches, the probabilistic outputs generated by the recognisers are combined assuming conditional independence, using the Bayes rule or some weighted lineal combination over the probabilities. For example, Kettebekov and Sharma (2000) used the architecture based on Multi-State Time Delay Neural Networks (MS-TDNN) shown in Figure 3.3 to recognise German letters.

In this architecture, the outputs of the acoustic and visual TDNNs are integrated in the combined layer, and the activation of each combined phone-state is the weighted sum of the activations of the units corresponding to the phonemes and visemes.[5] In the final layer, which copies the activations from the combined layer, a Dynamic Time Warping algorithm is applied in order to find the best path in the phoneme hypotheses that corresponds to a sequence of letter models.

---

[4] Audio and visual modalities are not totally synchronous, since the movements of the articulatory organs can start before or after the emission of the sound. This is known as the anticipation/retention phenomenon (Abry and Lallouache 1991).

[5] A viseme can be defined as the smallest visually distinguishable unit of speech.

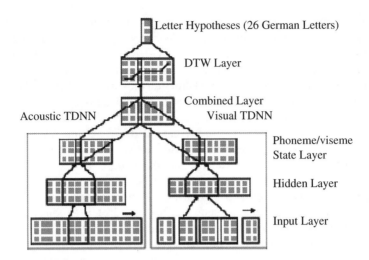

Figure 3.3 MS-TDNN architecture for AVASR (reproduced from Waibel et al. 1996, with kind permission of Springer Science and Business Media)

3.2.2.2 Fusion of Speech and Gestures

Speech and gestures are learnt during the first years of life and probably are the most natural communication means for human beings. Both modalities are used in MMDSs to increase the communication bandwidth and the interaction naturalness. The combination of both modalities allows, for example, a user to make a 2D 'deletion' gesture on a touch-sensitive screen while he utters the name of the object to delete. However, in comparison with speech, gestures were little investigated until the appearance of devices such as data gloves, which makes it possible to study the gestures made by the hands and track their movement in the 3D space. Initial research focused on obtrusive devices (e.g. data glove, pens, etc.) to include gesture information, which led to very restricted and unnatural multimodal interactions. Current technology uses artificial vision techniques to recognise hand movements in a less restricted way, which is very important to enhance the naturalness in the interaction with a MMDS.

In a multimodal interaction gestures allow the user to express ideas while he is using another input modality (e.g. speech) to interact with the system. However, to achieve a great level of cooperation between both input modalities, the interpretation of the data must be accurate and fast enough to allow the correct feedback to be provided to the user. It is very important to take into account the temporal relationship between gestures and speech in order to understand the meaning of both input modalities in a multimodal expression. According to the *semantic synchrony* stated by McNeill (1992), speech and gestures refer to the same idea, providing complementary information when both types of information occur simultaneously. Sharma et al. (2000) indicated that approximately 85% of the times a meaningful gesture is made, this gesture is accompanied by a keyword that is spoken during or after the gesture. So that, given a gesture input, a critical issue is to decide how much time to wait for a speech input (or vice versa) before considering the gesture as an unimodal interaction. To deal with this problem, the MATCH system (Johnston et al. 2002) uses a strategy based on timers that are kept very short by making them conditional on the activity in the other input modality. The system's fusion module is signalled in three cases: (1) when the user presses a 'click-to-speak' button; (2) when a speech or gesture recognition result arrives; and (3) when the user starts or finishes interacting with the graphic display. Using this mechanism, when a recognised sentence arrives, the system checks the status of the user. If he is interacting with the display, the fusion module waits for a recognised gesture to arrive, or otherwise waits for a short time and then considers the input to be unimodal. When a recognised gesture arrives at the fusion module, the procedure is analogous: the system checks the user status, if he has pressed the 'click-to-speak' button then waits for the speech recogniser output to arrive, or otherwise waits a brief time and considers the gesture unimodal.

In speech- and gesture-based systems it is common to use separate recognisers to deal with each modality, and combine at a higher level the corresponding outputs characterised by a timestamp. For example, in the QuickSet system (Cohen et al. 1997) the integration of these modalities is based on time restrictions: the events of both modalities are combined if they occur within a time window of 3 to 4 seconds. This approach has also been used by Holzapfel et al. (2004), who used fusion of speech and gestures in a multimodal system designed to be used in a human-like robot which helps in a kitchen scenario (e.g. to bring dishes into the kitchen, put them into the dishwasher, turn on lights, etc.). The author clearly shows the benefit of making the fusion of both modalities, given the changing environment conditions in terms of noise and lightning, which complicates the ASR and the visual tracking of the user. The multimodal inputs are stored using TFSs (studied in Section 2.2.2) and the fusion is carried out at the semantic level (studied in Section 2.2.1) using a set of rules that are applied to the input multimodal data. The left-hand side of the rules defines how the information chunks are integrated, whereas the right-hand side defines preconditions and restrictions that determine whether the rules are applicable (taking into account the number, modality, time properties, semantic content of the information chunks, etc.). The arrival of the different information chunks at the data fusion module varies depending on the delays caused by each recogniser. Thus, similar to the method used by Cohen et al. (1997) and Johnston et al. (2002), the fusion module developed by Holzapfel et al. (2004) waits for a determined time for the gesture information to arrive, but only if the content of the spoken input chunk suggests that there is information to disambiguate. The fusion of both modalities reduces the rate of falsely detected gestures from 53% to 26%, for a test set of 102 multimodal inputs.

3.2.3 Architectures of Multimodal Systems

Several architectures for SDSs and MMDSs have been proposed in the past decades. In the case of SDSs, a widely used architecture is called *Galaxy Communicator* (Seneff et al. 1998), which is a distributed, message-based, hub-centred architecture that has been used to set up, among others, the MIT's Voyager and Jupiter systems (Glass et al. 1995; Zue et al. 2000) (briefly discussed in Section 4.2). In the case of MMDSs, a design principle seems to be that the architecture should not only process information but also its meaning, taking into account the contextual information available when the fusion of the diverse information chunks is performed (Binot et al. 1990). The *unification* and *overlay* operations discussed in Section 5.4.4 are methods to implement this principle. Also, according to Hill et al. (1992), a MMDS must allow the user to combine the interaction modalities at any moment, the dialogue history must always be accessible (on-line and when the interaction has finished), and the system must be open to further enhancement without being completely re-implemented.

Another architectural principle seems to be the use of a common format to represent the information chunks provided by the different recognisers and processing modules. This principle is implemented by using methods to store multimodal information (Section 2.2.2). For this purpose, in addition to classical methods (e.g. frames or TFSs), there is a variety of XML-based languages such as MMIL (Multimodal Interface Language), M3L (Multi-modal Markup Language) or EMMA (Extensible Multimodal Annotation Markup Language). MMIL is a result of the IST MIAMM[6] European project whose main features are the management of salience, the status of secondary events in user utterances and the status of speech acts. M3L is a language developed in the SmartKom project to represent the information provided by the different modules of the SmartKom system (speech recogniser, gesture recogniser, face detection, fusion of multimodal information, etc.). The language represents all the information regarding synchronisation as well as confidence scores on the recognised hypotheses. However, contrary to MMIL, M3L does not have a meta-model and its syntax is inflexible. Its main advantage is its great potential to represent information using a set of XML schemas that represent word, gesture and facial expression hypotheses, as well as the results of the fusion module and the information to provide to the user (Wahlster et al. 2001). EMMA[7] is the result of the efforts made by the W3C working group to facilitate multimodal access to the Web. As in the previous cases, the language is based on XML and has been developed to represent the information chunks generated by the diverse modules of MMDSs: recognisers (speech, handwriting, etc.), fusion, dialogue management and response generation.

3.2.3.1 Open Agent Architecture (OAA)

The Open Agent Architecture (OAA) has been developed by SRI International to facilitate the setting up of applications using a variety of autonomous and distributed *agents*. These agents are software processes that may be written in different languages (e.g. C, Prolog, Lisp, Java, Visual Basic, Delphi, etc.), for a range of platforms (e.g. Windows, Solaris, etc.) and to interface with other agents using the Interagent Communication Language (ICL) (Moran et al. 1997). The architecture adds new agents or replaces agents in execution time, which are immediately available for the other agents. Another feature of the architecture is that the applications can be executed on a low powerful computer (e.g. a PDA), since the user interface agents executed on that computer can access other agents running in other, more powerful computers.

The core of the architecture is the *Facilitator* agent, which is responsible for the cooperation and communication of the agents. When an agent is added to an application, it registers its services in the Facilitator. Thus, when a particular agent requires the work of other agents in the application, it sends a request to the Facilitator which delegates the work to the agents that have registered as able to carry out the required task. The ability of the Facilitator to handle complex requests of the agents is an important element of the OAA architecture. Its goal is to minimise the information and the assumptions the system developer must embed in the agents, which facilitates reusing them for other applications.

---

[6] http://www.miamm.org
[7] http://www.w3.org/TR/emma/

Using this architecture, the different modules of a MMDS (Figure 1.1) are implemented as agents. The architecture generally includes an agent to coordinate the modalities and another to handle the user interface. The modality coordination agent is responsible for combining the information provided by the diverse input modalities, and generates a simple representation that captures his intention. The task of this agent is to resolve references, obtain the data not provided in the user requests, and resolve the possible ambiguities using the context, equivalencies or redundancies.

The user interface agent handles the diverse input modalities, which allows applications be isolated from the details of the modalities used in the interaction, thus notably simplifying the addition of new interaction modalities. When this agent detects a user input, it invokes the recogniser that must handle it and sends its output to another agent that obtains the user intention in a given representation. For example, if the input is provided via speech, this agent sends the voice samples to the ASR agent and then sends the recognised sentence to the NLP agent. The obtained representation is then sent to the facilitator agent, which sends it to the dialogue management agent to decide the next system action.

Considering its advantages, several researchers have used this architecture to set up a variety of MMDSs, for example, Moran et al. (1997) to set up a map-based tourist information application, McGee et al. (1998) to set up the QuickSet system, Cheyer and Julia (1999) to set up a multimodal map application, Johnston et al. (2002) to set up the MATCH system to provide restaurant and subway information, and Lemon et al. (2001) to implement the WITAS system to communicate with autonomous mobile robots. For example, in the latter system, the ASR is carried out by an agent based on a Nuance speech recognition server, the NLP is performed by an analyser based on the SRI Gemini parser, and the speech synthesis is carried out by an agent based on the Festival speech synthesis system.[8]

The OAA architecture is also used in the Chameleon platform, a general suite of tools in the form of a software and hardware platform that can be adapted to conducting intelligent multimodal interfaces in several application domains (Brøndsted et al. 2001). The architecture includes several agents: blackboard, dialogue manager, domain model, gesture recogniser, laser system, microphone array, speech recogniser, speech synthesiser, natural language processor, and a module called 'Topsy' that represents knowledge and behaviour in the form of hierarchical relationships.

3.2.3.2 Blackboard

The blackboard architecture stems from the field of Artificial Intelligence. Its goal is to share information between a heterogeneous set of agents that collaborate to resolve problems. The name 'blackboard' denotes the metaphor of a group of experts working collaboratively around a blackboard to resolve a complex problem. The blackboard is used as the repository of shared information that consists of the deductions and suppositions handled by the system during the problem resolution. Each expert observes the information on the blackboard and tries to contribute to the resolution using its own strategy. A facilitator agent controls the chalk (shared resource) acting as an intermediary among the experts competing to write on the blackboard, and decides at every moment the agent that must provide information to the

---

[8] http://www.cstr.ed.ac.uk/projects/festival/

blackboard, considering the relevance of its contribution. This architecture has been used by several researchers to develop dialogue systems. For example, Wasinger et al. (2003) used it to represent, analyse and make the fusion of multimodal data in a mobile pedestrian indoor/outdoor navigation system set up in a PDA device. The system architecture fusions speech with two gesture types: pointing to objects with a stylus on the PDA display, and pointing to objects in the real world. In order to process this second type of gesture, the system uses information concerned with the position, direction and speed of the user provided by the compass and GPS units incorporated into the PDA. This way, the user can make references to objects in the real world and the system can determine to which object the user is referring. For example, he can say 'What is this?' or 'Take me to here' while he points on the display, or 'Tell me about that church over there' while he points to it in the real world with this hand. Table 3.1 shows some blackboard entries from the recognisers (speech and gesture) and sensors (compass and GPS unit). Each entry has a time stamp (in milliseconds) indicating at what time the input was generated, a confidence score provided by the corresponding recogniser, and a value or a pointer to the corresponding object. To carry out the fusion, the system carries out either *unification* or *overlay* operations (discussed in Section 5.4.4) to integrate the entries in the blackboard.

As a variant of the previous architecture, Alexandersson and Becker (2001) used a multi-blackboard architecture in which the different components communicate to each other writing and reading in the so-called *information data pools*. This architecture, used to set up the SmartKom system, is shown in the Figure 3.4 (Pfleger et al. 2002).

In this architecture, the discourse modelling (a submodule of the dialogue manager shown in Figure 1.1) receives the hypotheses generated by the intention recognition module. The hypotheses are validated and enriched with information about the dialogue history, and a score is computed that indicates how well the hypotheses fit the dialogue history. Taking this value into consideration as well as the scores provided by the other recognition modules, the intention recognition selects the best hypothesis. In order to enrich a hypothesis with information about the dialogue history, the hypothesis is compared with segments of the previous dialogue history, using the unification and overlay operations (Section 5.4.4).

3.2.4 Animated Agents

In this section we focus on another representative feature of current MMDSs: the so-called *animated agents* (also termed *conversational agents*, *talking agents* or *embodied agents*).

Table 3.1 Inputs as written to the media fusion blackboard

Source type	MI	MI type	Unique ID	Time	Conf. value	Modality ptr. or value
PDA	Sensor	Direction	100	1046788415000	–	directionVal
PDA	Sensor	Velocity	101	1046788415000	–	velocityVal
PDA	Speech	Microphone	102	1046788415000	0.8	*speechPtr
PDA	Gesture	Stylus	103	1046788415000	0.9	*objectPtr

Source: Reproduced by permission of Wasinger et al. 2003.

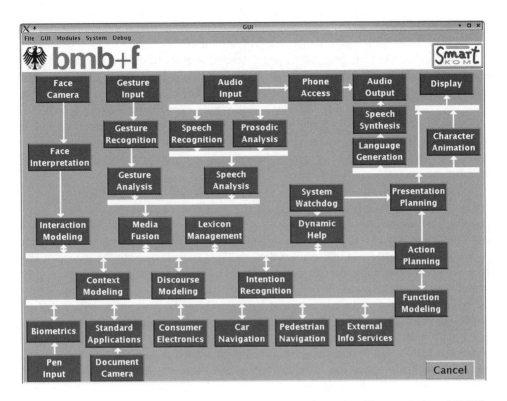

Figure 3.4   The architecture of the Smartkom system V2.1 (reproduced by permission of DFKI)

The most sophisticated agents have the form of human-like characters, either realistic such as the 'Rachel' agent of the CSLU toolkit (discussed in Section 6.2.3), or cartoon-like such as the 'Gurney', 'Lilly' or 'Percy' agents of the same toolkit (Ma et al. 2002), or the agent of the Rea system (Bickmore and Cassell 2004). Less sophisticated agents take on a variety of real or imaginary characters, e.g. the Smartakus agent of the SmartKom system (Pfleger et al. 2003), the agent of the Olga system (Beskow 1997), or the 'Rowdy' or 'Marge' agents of the CSLU toolkit. Animated agents are shown on a computer screen and (with more or less sophistication) are able to move the lips and make facial expressions and/or body gestures in synchronisation with the speech output. They are used in dialogue systems taking into account that the human face (and in general the human body) is a very important communication means. In fact, facial expressions and oral channels support and complement each other (e.g. the visual channel may clarify and disambiguate what is being said). Moreover, human beings not only move their lips while speaking but also raise their eyebrows, make expressive wrinkles or head nods to emphasise words, move their eyes, blink, etc. Researchers in several fields such as ethology, psychology, speech therapy, interpersonal communication and design of human–machine interfaces, indicate that facial expressions can enhance communication since they transmit emotions, emphasise some words or sentences, support the interaction in dialogues, transmit the relevant information regarding the discourse structure, and help in the turn-taking protocols (Pelachaud et al. 1993; Henton and Litwinowicz 1994). Moreover, facial expressions can transmit the speaker attitude, replace sequences of words or accompany them, and be useful to disambiguate what is being said when the acoustic signal is degraded by noise.

Taking into account all these advantages, several authors have developed methods to integrate appropriate intonation and facial expressions in order to enhance the intelligibility and naturalness of the responses generated by MMDSs. For example, Pelachaud and Prevost (1995) developed a system that integrates contextually appropriate intonation with facial expressions, and generates 3D facial animations automatically from the semantics and discourse representation. The intonation model is based on a combinatory categorical grammar (Steedman 1991) which easily integrates the syntactic constituency, prosodic phrasing and information structure.

A very important factor in the success of an animated agent is the synchronisation of the oral and visual output channels (Guiard-Marigny et al. 1996). The reason is that facial expression changes continuously, and many of these changes are synchronous with what is being said and with the effect the speaker wants to transmit (e.g. smile when remembering a happy event) or with the speech of the conversation partner. Thus, increasing the animated agent intelligibility requires controlling the facial animation articulatory parameters to make speech visually and acoustically perceived as generated by the movement of the articulatory organs. The synchronisation is generally achieved by using the audio channel as 'synchronisation clock'. For example, in House et al. (2001) for each phoneme the audio component sends a signal to the video component, which calculates the corresponding position of the oral cavity to visualise the viseme. In order to obtain correctly simulated output it is very important that the key phonemes are clearly visualised; for example, when visualising the /p/ phoneme it is very important that the animated agent clearly closes the lips. Beskow (1995) dealt with the synchronisation problem using a set of rules that take into account the coarticulation effects. Using these rules, gestures and facial expressions are generated using procedures that control the joined update of diverse parameters (e.g. time, value, direction, duration, etc.), allowing the agent to make complex gestures that require the simultaneous update of a great number of parameters. In addition to temporal coherence, there must also be coherence between the information transmitted visually and the acoustic information supposedly associated with the visual information. Thus, an animated agent can enhance the intelligibility of a speech synthesiser if its facial movements are coherent with the acoustic information supposedly generated by these movements. In this condition is not observed, the animated agent may definitely degrade the intelligibility of the spoken messages generated by the synthesiser.

Several approaches have been proposed to carry out the facial animation of animated agents, ranging from pre-recorded images to real simulation of physical movements. The initial human-like animated agents were based on the digitalisation and interpolation of images taken from real people making different expressions. Although this technique obtains good results, adding a new expression requires creating a new model and digitalising it, which takes time. In order to increase the naturalness and sensation of realism, more sophisticated approaches have been presented; in this section we only describe briefly the parametric and physiological models.

Parametric models are based on form parameters (related to the position and size of nose, eyes, etc.) and expressive parameters (related to the movement of eyebrows, mouth, eyelids, etc.). The agent animation is carried out by changing the values of the corresponding parameters as a function of time. This approach has been used by several authors, for example, Parke (1982), Hill et al. (1998) and Nahas et al. (1998). Parke (1982) presented a facial animation method that only pays attention to the external surface of the face without considering the muscles, and uses a set of parameters to deform the surface in order to create the desired facial expression. This method is simple, has small computational requirements and is very appropriate for lip animation; however, its main disadvantage is that it does not allow modelling the movement propagation. The values for the parameters can be obtained in a relative easy way by tracking the movement of important points of a real human face; for example, Massaro (2002) measured the movement of 19 points, as can be observed in Figure 3.5. Most parametric models are based on the FACS system (studied in Section 3.2.4.1), which uses the movement of facial muscles to specify and define the range of parameters.

Physiologic models can model the skin properties and muscle actions very well since they are based on biological studies and thus allow for very real-like animation, but a higher computational cost. These models can be divided into three groups: (1) *structural*; (2) based on *muscles*; and (3) based on *facial skin*. In the structural models the face is modelled using a hierarchy of regions (e.g. forehead, lips, chin, etc.) and subregions (e.g. upper lip, lower lip, etc.); the regions can either be shortened by the effect of muscles or affected by the movement of adjacent regions (Platt and Badler 1981). The models based on muscles combine muscle anatomic features with face properties (e.g. skin elasticity); each muscle is represented as a vector that indicates a direction, a magnitude and an influence area (Pelachaud et al. 1993). These models tend to be computationally expensive, which is an important drawback given the required synchronisation with the speech output. The models based on facial skin consider that the skin is made of different-thickness surfaces which are modelled using matrices (Terzopoulos and Waters 1990). These models can provide very realistic results but their main disadvantage is that it can be complicated finding the activation level for a determined facial expression.

3.2.4.1 FACS

FACS (Facial Action Coding System) is a comprehensive, notational system developed by the psychologists Ekman and Friesen (1978) to objectively describe facial activity. The

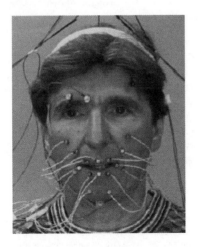

Figure 3.5   Tracking facial movements while speaking (reproduced by permission of Massaro 2002)

system is based on facial anatomic studies that consider facial animation as a consequence of muscle activity (Ekman and Rosenberg 1997). Thus, it describes visible facial expressions (either emotional or conversational) at the muscle level. The basic elements of the system are the so-called *action units* (AU), which represent the contraction of a muscle (or of a set of inter-related muscles). Table 3.2 shows a description of AUs.

The combination of AUs generates more than 7,000 facial expressions. For example, Table 3.3 shows a possible[9] configuration for the six basic expressions (surprise, fear, disgust, anger, happiness and sadness) shown in Figure 3.6, which includes a 'neutral' expression.

Table 3.2   List of action units (AUs)

AU	Description
AU1	Inner Brow Raiser
AU2	Outer Brow Raiser
AU4	Brow Lowerer
AU5	Upper Lid Raiser
AU6	Cheek Raiser
AU7	Lid Tightener
AU8	Lips Toward Each other
AU9	Nose Wrinkler
AU10	Upper Lip Raiser
AU11	Nasolabial Deepener
AU12	Lip Corner Puller
AU13	Cheek buffer
AU14	Dimpler
AU15	Lip Corner Depressor
AU16	Lower Lip Depressor
AU17	Chin Raiser
AU18	Lip Puckerer
AU19	Tongue Show
AU20	Lip Stretcher
AU21	Neck Tightener
AU22	Lip Funneler
AU23	Lip Tightener
AU24	Lip Pressor
AU25	Lips Part
AU26	Jaw Drop
AU27	Mouth Stretch
AU28	Lip Suck
AU29	Jaw Thrust

---

[9] There are many variations of these combinations to obtain the expressions, each having a different intensity of actuation. For example, the 'happiness' expression can also be considered to be a combination of pulling lip corners (AU12, AU13) and/or mouth opening (AU25, AU27) with upper lip raiser (AU10) and a bit of furrow deepening (AU11).

Table 3.2 (*continued*)

AU	Description
AU30	Jaw Sideways
AU31	Jaw Clencher
AU32	Lip Bite
AU33	Cheek Blow
AU34	Cheek Puff
AU35	Cheek Suck
AU36	Tongue Bulge
AU37	Lip Wipe
AU38	Nostril Dilator
AU39	Nostril Compressor
AU41	Lip Droop
AU42	Slit
AU43	Eyes Closed
AU44	Squint
AU45	Blink
AU46	Wink
AU51	Head turn left
AU52	Head turn right
AU53	Head up
AU54	Head down
AU55	Head tilt left
AU56	Head tilt right
AU57	Head Forward
AU58	Head back
AU61	Eyes turn left
AU62	Eyes turn right
AU63	Eyes up
AU64	Eyes down
AU65	Wall-eye
AU66	Cross-eye

Table 3.3 Combination of AUs to obtain basic facial expressions

Basic expression	Set of AUs
Surprise	AU1, 2, 5, 15, 16, 20, 26
Fear	AU1, 2, 4, 5, 7, 15, 20, 25
Disgust	AU4, 9, 10, 17
Anger	AU2, 4, 5, 10, 20, 24
Happiness	AU6, 11, 12, 25
Sadness	AU1, 4, 7, 15

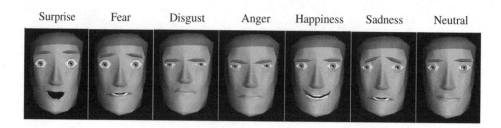

Surprise   Fear   Disgust   Anger   Happiness   Sadness   Neutral

Figure 3.6 The six basic facial expressions (and a 'neutral' expression) (reproduced by permission of Fabri et al. 2002)

It is possible to combine the effects of AUs by addition, dominance (an AU disappears in favour of other AU), or substitution (an AU is eliminated when other AUs make the same effect). In addition to create facial expressions, there are AUs to facilitate the control of the head movements (e.g. turning, tilting and nodding). Some AUs can operate on either side of the face independently of the other side, in which case the user must specify 'Left', 'Right' or 'Both'. The independent control of both sides of the face makes it possible to create asymmetric effects that are important for some expressions, such as winking and various other idiosyncrasies. Although in the AU actuation several muscles are involved, each AU has associated an intensity parameter that must be adjusted carefully. The great number of AUs defined in the FACS system makes it tricky to create and edit facial animations. Moreover, AUs are difficult to be manipulated by non-trained users. Because of these reasons, several FACS-based animation tools have been developed, for example, FACES (Facial Animation, Construction and Editing System) (Patel and Willis 1991).

3.2.4.2 MPEG-4

MPEG-1 and MPEG-2 are well known and widely used standards for video storage and distribution. MPEG-4 is a standard for multimedia information compression that is having a great impact on a variety of electronic products. It is the first standard based on visual objects, contrary to most video representation standards that are based on frames. MPEG-4 carries out the facial animation by composing audiovisual objects with specific behaviours, especially in terms of time and space. The approach for composing the objects allows new functions, such as the efficient coding of each data type, which is very important for critical bandwidth conditions. Another very powerful feature is the combination of real and synthetic content. The animation of 3D facial models requires animation data that can be synthetically generated or extracted from the analysis of real faces, depending on the application. The face model defined in MPEG-4 is a representation of a human face that makes intelligible the speech visual manifestations, transmits emotional information through the facial expressions, and reproduces the speaker face as faithfully as possible. The standard uses two types of parameter: Facial Animation Parameters (FAPs) and Facial Definition Parameters (FDPs). The FDPs are used to configure the 3D facial model that will be used, adapting a previously existing or using a new one.

The animation of the new model (or the adapted one) is carried out using the FAPs. To create a face, MPEG-4 defines 84 feature points (FPs) located in a face model that describes a

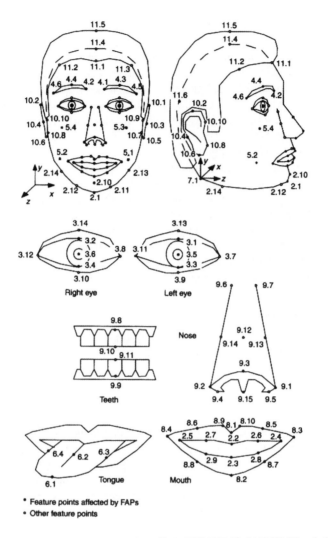

Figure 3.7 MPEG-4 feature points (from ISO/IEC IS 14496/2 Visual, 1999)

standard face (Figure 3.7). These points are used to define the FAPs and calibrate the facial models when different face players are used. The FAPs encode the facial animation by reproducing expressions, emotions and word pronunciations. They generate a complete set of basic facial animations, as well as representing most facial expressions. The set of parameters contains two high level parameters: *viseme* and *expression*. The viseme parameter represents visemes in the face without the need to express them in terms of other parameters. To allow speech coarticulation and face movements, a transition is made from a viseme to another fusing both using a weight factor. The expression parameter defines six high level expressions: joy, sadness, anger, fear, disgust and surprise. Contrary to what happens with visemes, facial expressions are generated using a value that defines the excitation of the expressions. It is possible to carry out the facial animation using these two FAPs but to obtain better results it is preferable to use the other (low-level) parameters. These parameters are related to face regions such as lips, bottom of chin, left corner of left eyebrow, etc. Each FAP corresponds with a FP and defines low-level deformations applicable to the FP with which it is associated. It is not relevant how this model is made available at the receiver. There are 68 FAPs divided into 10 groups, including parameters for head movements (e.g. head pitch and jaw angles) and in-face movements (e.g. opening of eyelids, opening of lips, movement of inner lip corners, etc.).

The FAPs represent a set of standard inputs the animator (application developer) can use; however, since they are low-level non-trivial parameters, it is generally preferable to use a tool that generates them from scripts written in languages such as VHML, APML or BEAT (see below), which integrate speech, emotions and facial expressions. The FAPs are model-independent parameters, thus in order to use them with a specific face model it is necessary to calibrate them. This process can be carried out using face animation parameter units (FAPUs), which are model-specific parameters that define fractions or distances between key facial features of the face model, for example, the separation between the eyes. Thus, FPs and FAPUs allow that the MPEG-4 compliant face players can be model-independent. Coding a face model using FPs and FAPUs allows developers can easily interchange the models without needing to consider about calibration or parameterisation for the animation (Balci 2004).

3.2.4.3 XML-Based Languages

Several XML-based languages have been developed to easily control the behaviour of animated agents using high-level primitives to specify behaviour acts similar to those made by human beings. In this section we comment briefly some of them, namely APML, DPML, HML, MPML, MURML and VHML:

- *APML and DPML.* In order to make an animated agent behave similar to a human being it is necessary to take into account that its verbal communication must be accompanied by non-verbal communication. Taking this fact into account, De Carolis et al. (2002b) proposed an animated agent's architecture based on two components: *mind* and *body*. The mind represents the personality and intelligence of the agent, which generates the emotional response to the events occurring in its environment. The body represents the physical appearance of the agent through which it interacts with the user using synchronised speech and facial expressions. The communication between both components is carried out through a *plan enricher* module that receives a communicative plan expressed in DPML (Discourse Plan Markup Language) generated by the mind, adds the semantic contents to communicate to the user, and obtains the input for the body expressed in APML (Affective Presentation Markup Language).
- *HML.* Another effort to help in specifying the animated agent behaviour at a high level is HML (Human Markup Language).[10] The goal of this language is to develop tools and repository systems on the Internet to enhance the fidelity of human communication. The language includes tags to specify several aspects concerned with the human communication in real situations, for example, cultural, social, physical or psychological.
- *MPML (Multimodal Presentation Markup Language)* is a language designed to carry out attractive, multimodal presentations using animated agents (Tsutsui et al. 2000).

---

[10] http://www.humanmarkup.org

- *MURML (Multimodal Utterance Representation Markup Language)* is a language designed to be used as an intermediary between planning and animation tasks to generate multimodal output using animated agents (Kranstedt et al. 2002). Its main advantage is the description of the gestural behaviour of the agent.
- *VHML (Virtual Human Markup Language)*[11] is a language designed to take into account HCI aspects concerned with facial and body animation, dialogue management, representation of emotional aspects, management of multimodal information and TTS conversion. It is based on other languages: EML (Emotion Markup Language), GML (Gesture Markup Language), SML (Speech Markup Language), FAML (Facial Animation Markup Language), BAML (Body Animation Markup Language), XHTML (eXtensible HyperText Markup Language) and DMML (Dialogue Manager Markup Language). Its goal is to facilitate the creation of animated agents as natural and realistic as possible to be used in Web pages or standalone applications, by means of tags that identify specific parts of the body (e.g. right eye), gestures and emotions (e.g. happy, sad, etc.).

3.2.5 Research Trends

To conclude this chapter, in this section we briefly discuss two research areas that are becoming increasingly important in the field of MMDSs: emotion, personality and social behaviour which aim at making the interaction with these systems more natural and similar to the one carried out between humans.

3.2.5.1 Emotional Dialogue

For a long time, the main paradigm in the research of dialogue systems was the information interchange considering exclusively the semantic content of the user interaction. In the last few years the research community has realised that emotions and other attitudes of interlocutors play a very important role in the interaction, which has attracted a lot of interest. The goal is that the automatic recognition of emotions can be used to allow dialogue systems to respond more appropriately to the user state of mind. For example, a computer could turn down the background music in a stressful situation for the user, or suggest a particular type of movie depending on his emotional state, etc. In many aspects, this research field represents the less explored frontier in HCI, probably because computers are generally considered logical and rational machines, incompatible with the apparently illogical and irrational nature of emotions. Another factor might be that, although the human ability to express and feel emotions is extraordinary, there is no general agreement about how to define them exactly. These reasons make the recognition of emotions a particularly difficult task.

The different approaches to the problem of emotion recognition are accompanied by different classification models. Some models consider emotions can be continuously represented in a multidimensional feature space; for example, a very used model is based on the arousal-valence[12] bidimensional space (Lang et al. 1990). Other models consider emotions as

---

[11] http://www.vhml.org

[12] Arousal refers to the intensity of the emotion (e.g. calm or aroused) while valence (also known as 'pleasure') refers to how positive or negative the emotion is (pleasant or unpleasant).

represented discretely; for instance, Ekman (1992) defined five, six or more distinguishable emotional states (e.g. happiness, sadness, anger, fear, surprise, disgust and neutral). A widely used discrete model is OCC (Ortony et al. 1988), which considers 22 discrete emotional states. This model explains emotions in terms of cognitive processes related to user intentions, standards, likes and dislikes regarding the actions of other agents, and the state of the world resulting from such actions. Examples of how this model has been used to describe and synthesise emotions can be found in André et al. (1999), Picard (2000) and Streit et al. (2004).

The idea behind using emotion recognition in dialogue systems is that the interaction can be more pleasant, comfortable and productive. The systems would provide help and assistance to a confused user and try to animate a frustrated user, thus reacting in a more appropriate manner than just ignoring the user state, which is the typical situation in most current dialogue systems. This idea is supported by several previous works indicating that emotion recognition may be useful to reduce user frustration (Klein et al. 2002; Prendinger et al. 2003). Moreover, given that emotions represent a key element of the social behaviour of human beings, dialogue systems able to recognise human emotions would be perceived as truly effective devices in real communication. Emotion recognition is considered an additional interaction channel that enhances the effectiveness of the communication. In the same way as gestures can serve to disambiguate the meaning of some sentences (e.g. put-that-there), emotions can be used to disambiguate some sentences, for example, when sarcasm is used (e.g. 'Excellent, that's exactly what I needed!').

Although there have been many advances in the engineering of affective systems to show human emotions in appropriate situations (e.g. using the animated agent's facial expressions, gestures and/or emotional synthesised speech), the recognition and interpretation of human emotions in dialogue systems are still in their infancy. Emotion can be inferred indirectly by monitoring system deficiencies, such as errors of the ASR module or the inaccessibility to information servers. Alternatively, it can be derived directly from the syntactic and semantic point of view (e.g. use of affective phrases), acoustic realisation (e.g. talking speed, volume, etc.), facial expressions, gestures and/or other physical and psychological signs, although the methods based on the user face and speech processing have been more explored (Healey and Picard 1998; Picard 2000). Recognising emotions requires using a device that captures signals from the user (e.g. microphone, bio-sensor or camera) and a classification scheme that uses features obtained from such signals to recognise the emotions. In this section we discuss the use of bio-sensors and cameras for artificial vision.

3.2.5.2 Bio-sensors

Emotion recognition based on bio-sensors provides several advantages in comparison with other methods, for example, providing great robustness against environmental conditions. Since the size of these sensors is becoming smaller day by day, it will be common to have them in cloth and jewellery articles (e.g. rings, bracelets, etc.); in fact, it is now possible in many cases (Healey and Picard 1998). On the contrary, recognition using cameras is difficult when there is little light and the user is moving. Using speech signals presents many problems since the user may be listening to music, watching a movie, talking to other people, etc. Moreover, the environmental factors discussed in Section 2.1.1 for ASR must be considered. Therefore, there are many arguments in favour of developing systems able to recognise the user emotional state from biological signals, which can also be combined with other methods (i.e. speech and facial expressions) to combine their respective advantages.

Some emotions cause a great variety of body reactions, usually called *bio-signals*, which can be detected by the so-called *bio-sensors*. For example, Haag et al. (2004) monitored the following bio-signals to determine the emotional state of the user: *electromyography*, to measure the activity or tension of a particular muscle; *electrodermal activity*, to measure the skin conductivity (which increases with sweat); body temperature; blood volume, to measure the amount of blood circulating through the veins (e.g. of a finger); electrocardiograms, to measure the heart contraction activity, and breathing, to measure how deep and fast the user is breathing. Using a combination of these measures, the authors obtain a set of features that are used to train an ANN and determine user emotional state in terms of arousal and valence. The experimental results show that the network obtains 96.6% and 89.9% in the recognition of both measures, respectively.

According to Lang (1995), there is a relation between physiological signals and arousal/valence given that the activation of the nervous system changes with emotions. Some authors have followed this idea to develop the so-called *affective dialogue systems*. For example, Prendinger et al. (2004) used two types of physiologic signals to set up an affective system that recognises user emotional state and also influences it by using an animated agent. On the one hand, the authors used a galvanic skin response signal (GSR) as an indicator of skin conductance, which increases linearly with the level of arousal. On the other hand, they used a electromyography signal (EMG) which measures the muscle activity and has been shown to be related to emotions of negative valence. In the experiments, the GSR sensors were connected to two fingers of the non-dominant hand of the users, whereas the EMG sensors were connected to the forearm of the same side of the body. The system processed in real time the physiological data obtained from the users (skin conductance and electromyography) and used the animated agent to provide affective feedback to the users (also called *empathic* or *sympathetic* feedback). The impact of the animated agent's affective response was measured comparing the skin conductance of users receiving the affective feedback against that of users who did not receive it. The agent affective response was as follows: on the one hand, it showed a concern expression if the user was detected to be aroused and with a negatively valenced emotion, and said, for example, 'I am sorry that you seem to feel a bit bad about this question.' On the other hand, the agent applauded the user and congratulated him if he was detected to be positively aroused, saying, for example, 'Well done!', 'Good job! You answered correctly.' The experiment results showed the affective behaviour of the system enhanced the interaction by making the users less frustrated if stressful events occurred during the interaction. When the agent generated empathic responses to frustrated users, the skin conductance was lower that when it did not generate this response. If the skin conductance is interpreted as signalling the stress or frustration level, if follows that the empathic feedback was useful to reduce the user negative reactions.

3.2.5.3 Cameras for Artificial Vision Combined with Other Modalities

A variety of approaches can be found in the literature to the problem of classifying facial expressions using artificial vision techniques. For example, Essa and Pentland (1995) considered the spatio-temporal energy of facial deformations from image sequences that define dense templates of expected motions, and used a Bayesian classifier to categorise the observed facial expressions according to the most similar average motion pattern. Bascle et al. (1998) tracked facial deformations using patterns generated from B-spline curves, and selected key-frames to represent basic facial expressions. Schweiger et al. (2004) used an ANN to integrate velocities (amount and direction of movement) in different sectors of the face.

Artificial vision techniques to recognise the emotional user state have been used by several MMDSs, mainly combined with other information sources. For example, in the SmartKom system, Adelhardt et al. (2003) used a method to identify four emotional states (neutral, angry, joyful and hesitant) based on a facial expression classification with eigenfaces, a prosodic classifier based on ANNs, and a discrete HMM for gesture analysis, all working in parallel. Also with the same dialogue system, Prendinger et al. (2004) used prosody combined with user facial expressions and information about problematic situations in the dialogue.

3.2.5.4 Personality and Social Behaviour

It is evident that non-verbal communication (e.g. facial expressions and gestures) plays a very important role in face-to-face dialogues between human beings. This communication type is normally used to transmit interpersonal attitudes and help understand the spoken messages. For example, an eyebrow rising can be used in a variety of circumstances to produce different communication effects. Thus, it is convenient to integrate this behaviour in MMDSs so that they can replicate more faithfully the behaviour of a real person in a given task. However, interacting with a dialogue system in the same way as with a human being requires that the system has its own *personality*. Several personality models can be found in the literature. One is the Five Factor Model (FFM) (McCrae and John 1992), which is a purely descriptive model with five dimensions: Extraversion, Agreeableness, Conscientiousness, Neuroticism and Openness. The model represents the traits of *extraversion* (social vs. misanthropic; outgoing vs. introverted; confidence vs. timidness), *agreeableness* (friendliness vs. indifference to others; docile vs. hostile nature; compliance vs. hostile non-compliance), and *neuroticism* (adjustment vs. anxiety; level of emotional stability; dependence vs. independence). Among others, André et al. (1999) used this model to set up the personality of the animated agents used in the Presence, Inhabited Market Place, and Puppet projects, focusing only on the *extraversion* and *agreeableness* dimensions, which, according to Isbister and Nass (1998), seem to be the most relevant for the social interaction.

Other authors have used Bayesian Belief Networks to represent the personality model, as, for example, Ball and Breese (2000), and Kshirsagar and Magnenat-Thalman (2002). The latter included an additional layer for the 'mood' feature, which filters the emergence of emotions that influence the interpretation of a situation. Thus, a person in a 'good' mood tends to have a positive interpretation of the situation which moderates the emotion he feels. Conversely, a person in a 'bad' mood has a negative interpretation of the situation, accentuating the felt emotion.

El Jed et al. (2004) represented the emotional state of an animated agent by a set of emotions whose intensities can change during the dialogue. Each emotional state at time t,

E_t, is an m-dimensional vector in which each element represents the intensity of an emotion represented by a value in the interval $[0,1]$:

$$E_t = \begin{pmatrix} e_1 \\ e_2 \\ \dots \\ e_m \end{pmatrix} \forall i \in [1, m], e_i \in [0, 1] \tag{3.1}$$

Adopting the emotional categories proposed by the OCC model, they consider four types of emotion: *satisfaction, disappointment, anger* and *fear* ($m = 4$):

$$E_t = \begin{pmatrix} e_{SATISFACTION} \\ e_{DISAPPOINTMENT} \\ e_{ANGER} \\ e_{FEAR} \end{pmatrix}$$

Therefore, for example, the state $E_t = (0, 0, 0.8, 0.3)$ means the agent is experiencing anger and a little fear. These emotions affect the state of the animated agent in several ways; for example, they can generate an affective reaction expressed by a facial animation that conveys the predominant emotion (e.g. fear), and a behavioural reaction expressed by a set of gestures or/and actions.

The authors represent the personality P as a static n-dimensional vector in which each element represents the dimension of the personality using a value in the interval $[0, 1]$:

$$P = (p_1, p_2, p_3, \dots, p_n) \; \forall i \in [1, n], p_i \in [0, 1] \tag{3.2}$$

As the FFM is one of the most accepted models of personality, they defined the personality focusing on the five dimensions of this model: *openness, conscientiousness, extravert, agreeableness* and *neurotic* ($n = 5$):

$$P = (p_{OPENNESS}, p_{CONSCIENTIOUSNESS}, p_{EXTRAVERT}, p_{AGREEABLENESS}, p_{NEUROTIC})$$

For example, the personality $P = (0.2, 0.8, 0.7, 0.5, 0.5)$ means the agent is very conscientious and extravert, not very open, and quite agreeable and neurotic. The authors consider that every emotion is influenced by one or more dimensions of the personality. The intensity of this influence is represented in the so-called MPE matrix ($m \times n$) by a value in the interval $[0,1]$, as follows:

$$MPE = \begin{pmatrix} \alpha_{11} & \alpha_{12} & \cdots & \alpha_{1n} \\ \alpha_{21} & \alpha_{22} & \cdots & \alpha_{2n} \\ \cdots & \cdots & \cdots & \cdots \\ \alpha_{m1} & \alpha_{m2} & \cdots & \alpha_{mn} \end{pmatrix} \tag{3.3}$$

where $\forall i \in [1, m], j \in [1, n], \alpha_{ij} \in [0, 1]$. An example of such an MPE matrix is the following:

$$MPE = \begin{pmatrix} 0 & 1 & 0 & 0.3 & 0 \\ 0 & 1 & 0 & 0.3 & 0 \\ 0 & 0 & 0.5 & 0 & 0 \\ 0 & 0 & 0.2 & 0 & 1 \end{pmatrix}$$

In this matrix, the first line (0, 1, 0, 0.3, 0) indicates that the emotion of *satisfaction* is strongly influenced by the *conscientious* dimension (value $= 1$) and lightly influenced by the *agreeableness* dimension (value $= 0.3$).

In human beings, the personality enables some social attitudes. For example, in western cultures (at least) it is usual to start conversations talking about neutral topics (e.g. weather conditions, travels, current news, etc.) before discussing the central points. In some sense, this initial phase of the conversation aims at knowing aspects of the personality and/or culture of the interlocutor, in order to define a style and a language for the communication that enhances the understanding of each other (Bickmore and Cassell 2004). Several studies have been made to check whether this social behaviour could also be applied to the interaction with dialogue systems. For example, Gustafson et al. (1999) carried out an experiment during the development of the August system to determine the type of sentences the users would say to the system's animated agent. They collected more than 10,000 sentences uttered by around 2,500 users. From the analysis of these sentences the authors observed that the users tried to socialise with the agent, using approximately one-third of the sentences. Reeves and Nass (1996) indicated that users prefer the systems that coincide with their own personality in terms of introversion or extroversion, whereas Bickmore and Cassell (2004) showed that the social behaviour notably affects the trust in the system for the extrovert users, while the trust is not affected for the introvert users.

The social feature of human dialogues has been implemented in several MMDSs, for example, Olga (Beskow 1997), PPP Persona (André et al. 1999), August (Gustafson et al. 1999) and Rea (Bickmore and Cassell 2004). These systems feature a personality specially designed to socialise with users in addition to interacting with them taking into account a specific task. For example, the Rea system mixes dialogue turns concerned with the task (real estate) with others specific to socialising with the user. Thus, in the task turns the system prompts for data concerned with the apartment to rent (e.g. number of bedrooms needed), whereas in the socialising turns it talks about the weather or tells stories about the lab in which it was developed, or about the real estate market. The system decides its next turn, taking into account that the task must be carried out in the most efficient way, and also that the turn must maximise the user trust in the system. Taking these factors into account, it decides whether its next turn must be concerned with the task or the socialisation, and in the second case, the type of socialisation turn.

3.3 Summary

In this chapter we studied MMDSs firstly addressing the benefits of such systems in terms of system input, dialogue management and system output. Next, we focused on development concerns, addressing initially the two classical development techniques (WOz and system-in-the-loop). Then we addressed the problem of data fusion, presenting the fusion of acoustic and visual information on the one hand, and that of speech and gestures, on the other. The system architecture, discussing two architectures commonly used for the setting up of multimodal systems: Open Agent Architecture and Blackboard, was outlined. Animated agents, discussing several approaches to facial animation (based on parametric and physiological models), as well as face and body animation standards and languages (FACS, MPEG-4 and several XML-based languages) were then introduced. To conclude, the chapter addressed research trends in MMDSs, discussing aspects concerned with the personality and the social behaviour.

3.4 Further Reading

Affective Dialogue Systems, Emotional Dialogue, Personality and Social Dialogue

- André, E., Dybkjær, L., Minker, W. (eds) 2004. *Affective Dialogue Systems: Tutorial and Research Workshop*, Springer-Verlag, Heidelberg.

Animated Agents

- Section 2.5 of Gibbon et al. (2000).

Architectures of MMDSs

- Part 2 of Taylor et al. (2000).

Data Fusion

- Sections 2.4 and 2.5 of Gibbon et al. (2000).

Multimodal Web-based Dialogue

- Chapter 11 of McTear (2004).

Survey of MMDSs

- Section 2.2 of Gibbon et al. (2000).

4

Multilingual Dialogue Systems

4.1 Implications of Multilinguality in the Architecture of Dialogue Systems

In this section, we discuss possible alternatives in constructing MLDSs and examine typical architectures. We start the discussion by investigating some alternatives in constructing these systems from the viewpoint of ASR, TTS, NLU, NLG, level of interlingua, and dialogue context dependency. Three possible approaches to the architecture are studied: interlingua, semantic-frame conversion and dialogue-control centred approach.

4.1.1 Consideration of Alternatives in Multilingual Dialogue Systems

To implement a MLDS, there are several design alternatives for the system components, from speech processing elements to application managers, and various ways of dividing the information processing in each component. The combination of these choices determines the architecture of the system. The main concern of these choices basically depends on which element assimilates the language differences and the task dependencies.

4.1.1.1 Speech Recognition

Many implemented MLDSs use a language-dependent speech recogniser to handle the speech input. Among other considerations, there is one important question in the set up of this module: the kind of language model that should be used. Basically, two approaches can be considered: rule-based and statistical. In the case of rule-based language modelling, which is usually written in BNF (Backus Naur Form) notation, multilingual grammars are typically written separately for each language, but they must be consistent in terms of the concepts they can represent. This approach is suitable for rapid system development but poses stricter restrictions on the user's input sentences. As a result, the dialogue generally becomes system-directed (Section 5.3.1). On the other hand, in the case of speech recognition with a statistical language model, the system can deal with spontaneous input to some extent. However, a

Spoken, Multilingual and Multimodal Dialogue Systems: Development and Assessment Ramón López-Cózar Delgado and Masahiro Araki © 2005 John Wiley & Sons, Ltd

drawback of this approach is that it needs a large amount of dialogue corpus obtained for the application domain and language. In many cases, it is difficult to collect enough data to prepare the language model for all the languages. Therefore, some kind of inter-lingual adaptation may be needed, for example, a task-dependent Japanese corpus can be created from the same task-dependent English corpus and then used to train the language model.

In addition, the ASR module for a MLDS needs the ability to define a pronunciation lexicon for foreign words, as it should deal with place names, people's names, restaurant names, etc. which can appear in many kinds of dialogue tasks and domains. In general, phonetic rules are written for the word dictionary in ordinary ASR systems. In the Voice Browser Activity[1] of the W3C, the Pronunciation Lexicon Specification (PLS)[2] is under development at the time of writing. This specification is intended to allow interoperable specification of pronunciation information for speech recogniser and synthesis engines. Using the independent phonetic alphabet IPA (International Phonetic Alphabet), PLS can specify the pronunciation for each lexicon, as shown in Figure 4.1.

As an alternative to this separation for each language, there is also the possibility to use one common recogniser for all the target languages (Ming Cheng et al. 2003). If the speech recogniser can deal with multiple languages, then a language identification subdialogue, which improves the usability of the system especially for naïve users, can be located at the beginning of the dialogue. This way the user can discover whether his native language can be used in the system, and if not available, then can easily switch to a second language within the same session (i.e. without hanging up and dialling another service number for interacting using another language). Another advantage of the multi-language recogniser is the memory-saving when used on small devices, such as cellular phones and PDAs. Although the recognition module is common to all the target languages, the recogniser output is a word sequence specific for each language. Therefore, modules that assimilate the language differences after speech recognition are needed.

4.1.1.2 Speech Synthesis

Unlike what happens with the speech recognition module, there is little problem with the speech synthesiser to generate the speech output. Also, there is little advantage in integrating the multi-language output into one module, except for the purpose of the above-mentioned memory-saving in small devices. However, at the level of dialogue description, the virtual

```
<?xml version="1.0" encoding="UTF-8"?>
<lexicon version="1.0" alphabet="ipa" xml:lang="en-US">
 <lexeme>
  <grapheme>Kyoto</grapheme>
  <phoneme>kjoʊtɔ</phoneme>
 </lexeme>
</lexicon>
```

Figure 4.1 An example of PLS

---

[1] http://www.w3.org/Voice/
[2] http://www.w3.org/TR/pronunciation-lexicon/

integration of all the target language increases the maintainability of the system. For example, the standardised markup language for speech synthesis SSML[3] (Speech Synthesis Markup Language) can use a 'lang' attribute that specifies the output language by an internationalised language code (for example, 'en-US' for US-English, 'ja' for Japanese, etc.). This attribute can be used in block elements, such as phrase (`<p>` element) and sentence (`<s>` element). A speech synthesiser conformed to SSML can pick up the expression indicated by the LANGUAGE parameter of the voice platform (for example, a VoiceXML[4] interpreter) to generate the appropriate output. An example of such an output is shown in Figure 4.2.

One important function required for the speech synthesis module in MLDSs is the ability to pronounce proper nouns in other languages (because of the same reasons discussed for the ASR module). Therefore, the speech synthesis module must interpret a language-independent phonetic description, such as IPA. The above-mentioned PLS is intended to be used in full-scale vocabulary management for both recognition and synthesis using such language-independent phonetic description. Also, using SSML, the `<phoneme>` element can provide a phonetic pronunciation for the text, as for example:

```
<phoneme alphabet="ipa" ph=" kjoʊtɔ">
   Kyoto
</phoneme>
```

4.1.1.3 Language Understanding

As shown in Figure 1.1, the basic architecture of a dialogue system has in its input interface an ASR module that converts the speech signal from the user voice into a word sequence, or a word lattice (or several sequences, in the case of N-best recognition, as discussed in Section 2.1.1). Also, it has an NLU (also called NLP) module that converts the output of the ASR module into a semantic representation. The most straightforward method for creating a MLDS is to prepare the ASR and NLU modules for each language, and make them generate a language-independent common semantic representation. In general, the implementation (or modification) of the NLU and NLG components can be a bottle-neck, as it is essential that the common semantic representation can be obtained from the equivalent surface expressions uttered in each language. Either using a grammar-based or a statistical-analysis method, the interlingua approach to multilinguality inevitably must face this problem.

```
<speak version="1.0" xmlns="http://www.w3.org/2001/10/synthesis"
    xmlns:xsi="http://www/w3/org/2001/XMLSchema-instance"
    xsi:schemaLocation="http://www.w3.org/2001/10/synthesis
    http://www.w3.org/TR/speech-synthesis/synthesis.xsd"
    xml:lang="en-US">
    <p>Thank you very much.</p>
    <p xml:lang="ja">ありがとうございました。</p>
</speak>
```

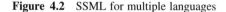

Figure 4.2 SSML for multiple languages

---

[3] http://www.w3.org/TR/speech-synthesis/
[4] http://www.w3.org/TR/voicexml 20/

As an alternative to the interlingua, it can also be used a direct mapping approach, which does not have an independent module for NLU and NLG for each language. In this approach, the result of the ASR module is directly mapped to the variables of the application-logic component, by using the mapping rules in the grammar for each language. For example, SRGS[5] (Speech Recognition Grammar Specification) makes use of such a mapping method. At the lexicon level (`<item>` element) or rule level (`<rule>` element), `<tag>` elements can specify the binding value for a variable when the corresponding word is recognised (or the corresponding rule is applied). Therefore, by making tag contents uniform, multilingual treatment can be achieved without a NLU component. Figure 4.3 shows an example of SRGS grammar.

4.1.1.4 Language Generation

In the case of MLDSs based on the interlingua approach (to be studied in Sections 4.1.2 and 4.2), the NLG component takes a common semantic representation as input, generates a word sequence, and passes it to the TTS subsystem. The problem of the NLG component is opposite to that of the NLU component, i.e. the NLG module must generate a consistent output for all the possible input interlingua representations, for all the target languages. This restriction may cause serious problems, especially in the system maintenance phase, since a slight modification of the input specification, requires the modification of the sets of rule for all the target languages.

In the case of the direct mapping approach for the NLG, a template-based sentence generation method carries out a direct mapping which does not require a sentence construction process. The role of the NLG component is to fill with data the slots of the pre-defined templates. This approach is useful for easy and simple tasks and domains, which allow a concept to be directly mapped to the task variable. However, it is not appropriate for more complicated tasks and domains, with sentences containing multiple dialogue acts or having modification relations between noun phrases. We will discuss how to lighten the generation in MLDSs in Section 4.3.

```
English Grammar
<rule id="rule1"scope="public">
  <one-of>
   <item> bus <tag>$.var="bus"</tag></item>
   <item> train <tag>$.var="train"</tag></item>
  </one-of>
</rule>

Japanese Grammar
<rule id="rule1"scope="public">
  <one-of>
   <item>バス<tag>$.var="bus"</tag></item>
   <item>電車<tag>$.var="train"</tag></item>
  </one-of>
</rule>
```

Figure 4.3 SRGS for multiple languages

---

[5] http://www.w3.org/TR/speech-grammar/

4.1.1.5 Level of interlingua

The abstraction level of the interlingua affects the component design and implementation. One extreme level is a task-independent representation which is similar to that used in Machine Translation (MT) systems. Another extreme level is a fully task-dependent representation which is almost equivalent to a command language for application control.

On the one hand, for task-independent semantic representations, previous research on MT can provide know-how about specifying language-independent semantic case frames. Basically, a task-independent meaning representation expresses a syntactic relation for each linguistic constituent, which allows that carefully constructed grammars can be reused for other tasks. However, as the meaning representation is based on syntactic information, it becomes difficult to ensure that sentences with the same meaning in various languages fall into the same interlingua representation. In addition, the dialogue manager tends to become more complex as it is responsible for interacting with the backend application following the input representation that might be far from application commands. Other possible drawback of this task-independent semantic representation is the lack of robustness, as construction largely depends on the rule-based syntactic analysis.

On the other hand, task-dependent semantic representations tend to be simple, and thus the burden of the dialogue manager is lightened. Many existing MLDSs use this approach. The grammar formalism used in the NLU is a kind of semantic grammar whose rules are written using semantic categories (e.g. date, time, person, place, etc.) instead of syntactic categories (e.g. noun, adjunctive, verb, etc.). As a natural consequence, this grammar is scarcely reused in another tasks or domains given that the semantic granularity differs for each task. For example, the *time* category must be finer in the schedule management domain than in the weather forecast domain.

The same discussion can be applied to the generation side of the dialogue system. A NLG module that receives task-independent semantic representations would be easily ported to another task or domain. However, since this task-independent semantic representation tends to be verbose, the semantic representation construction becomes more complex. On the contrary, a NLG module that receives task-dependent semantic representations requires task-specific rules to generate fluent expressions. Therefore, it is more difficult to reuse for other tasks or domains.

4.1.1.6 Dialogue Context Dependency

The treatment of the dialogue context represents another decision point for the system designer. In general, mixed- or user-initiative dialogue systems have to deal with spontaneous speech phenomena such as hesitations, disfluencies, fragmented sentences and indirect speech acts. In order to deal with these phenomena, the systems need to preserve the dialogue context. A possible way to deal with it is to use rich context information, while another is to try to exclude the phenomena by the system design, using a system-directed interaction strategy that limits the possible user expressions.

In the case of full treatment of the dialogue context, a language-dependent component is needed to manage the context update. For example, in the case of a system that handles pronoun resolution by a centring mechanism (Grosz et al. 1995), the dialogue context manager must maintain the list of **Cf** (forward-looking centres) ranking. The order update of this ranking list is based on several principles that are language-dependent, such as the syntactic preference (subject > object > other) is used in English.

On the contrary, in the case of exclusion of the dialogue context, a careful design of the dialogue flow is needed in order to exclude the appearance of user's ambiguous fragmented sentences that may have several possible interpretations. Despite the endeavour in the careful design of the dialogue flow, in many tasks, referring expressions may inevitably appear, such as 'I prefer the previous train', in which case the dialogue manager must resolve the problem (i.e. find the referent) by using the dialogue history constructed from the interlingua input.

4.1.2 Interlingua Approach

Following the architecture of the MMDS shown in Figure 1.1, an interlingua-based architecture is a straightforward approach to setup a MLDS. This architecture uses language-dependent NLU and NLG, task-dependent interlingua, and no context dependency. Table 4.1 shows possible choices for setting up this approach, while Figure 4.4 shows the structure of the resultant MLDS.

The ASR module receives a speech signal from the user and outputs a word sequence (or word lattice) hypothesis (or several in the case of N-best recognition) for the NLU module. The NLU module creates a semantic frame represented in a task-dependent interlingua after

Table 4.1 Choices for the interlingua approach

	ASR	NLU	Interlingua	Context	NLG
Choice	Grammar-based LM or N-gram	Syntactic and semantic analysis	Task oriented	N.A.	Rule-based or template-based
Language portability	Medium	Bad	Excellent	–	Bad
Task portability	Medium	Bad	Medium	–	Bad

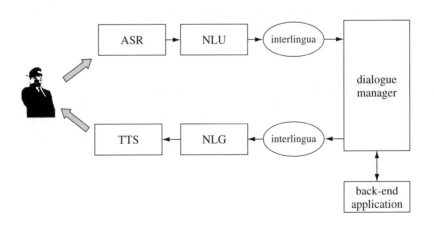

Figure 4.4 MLDS architecture based on interlingua

a syntactic and/or semantic analysis. The language differences are assimilated in this NLU module. Then, the dialogue manager interprets a task-dependent semantic representation, decides what to do next (e.g. ask the user for clarification, retrieve information from the back-end application, report results, etc.) and outputs a task-dependent interlingua for the NLG module. This module converts the interlingua into sentences expressed in natural language that represent either the result for the user's request, a status report from the system, or a system's clarification question. The language differences are also assimilated in this module. The obtained natural language expression is passed to the TTS subsystem to generate the system speech output.

The advantage of this architecture is its simplicity. The role of each component is easily decided according to its input and output specifications. In addition, once the specification of the task-dependent interlingua is fixed, the NLU and NLG components for each language and the dialogue manager can be developed independently.

The disadvantage is that this architecture assumes implicitly there is no interaction between the NLU component and the dialogue manager, except for the delivery of the interlingua. Some dialogue systems provide contextual information from the dialogue manager to the NLU and the ASR modules in order to restrict the search space in the ASR, and resolve anaphora and ellipsis in the NLU. If the system developer wants to deal with these contextual problems using this architecture, all the discourse context processing should be implemented in the dialogue manager using interlingua information only. The problem is that, in general, it is difficult to resolve such contextual problems without using language-specific knowledge.

Several MLDSs are based on this architecture, including Jaspis (Turunen and Hakulinen 2000, 2003), AthosMail (Turunen et al. 2004a, 2004b) and ISIS (Meng et al. 2002).

4.1.3 Semantic Frame Conversion Approach

One possible derivation of the interlingua-based architecture is a semantic frame conversion architecture which combines task-independent semantic analysis with task-dependent case frame conversion. This architecture consists of language-dependent ASR and TTS, two stages of NLU and NLG (one task-independent and the other task-dependent), and some kind of discourse level processing at the task-independent language module. Table 4.2 shows possible choices for implementing this approach, while the overall architecture is shown in Figure 4.5.

Table 4.2 Choices for the semantic frame conversion approach

	ASR	NLU	Interlingua	Context	NLG
Choice	Grammar-based LM or N-gram	Syntactic and semantic analysis	Two levels: Task independent and Task dependent	Language dependent discourse processing	Rule-based or template-based
Language portability	Medium	Medium	Good	Bad	Medium
Task portability	Medium	Good	Good	Bad	Good

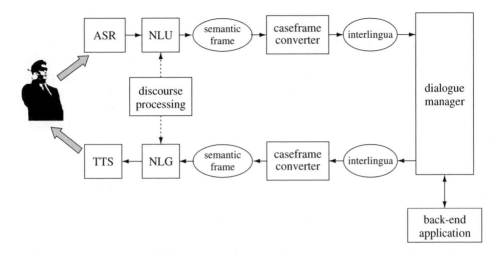

Figure 4.5 MLDS architecture based on semantic frame conversion

The problem with the interlingua approach is the setting of the level of interlingua, as discussed in Section 4.1.1.5. If the level of interlingua is too general, the dialogue manager becomes complex, while if it is too specific to the task and domain, it becomes difficult to port the NLU to another task/domain.

The semantic frame conversion approach carries out two processing steps for both the understanding and generation processes. In the understanding processing (upper part of Figure 4.5), the first step is the task-independent and language-dependent NLU. This step uses language-dependent syntactic knowledge and general purpose semantic knowledge for each language, which can be re-usable for other tasks and domains. The second step converts the task-independent semantic frame (which is the output of the NLU module) into a task-dependent and language-independent semantic frame that can be directly converted into an application command (e.g. in SQL).

In the NLG component (lower part of Figure 4.5), the first step is to convert a task-dependent response frame (e.g. a retrieved record from the database) into a semantic frame suitable for generating a response in each target language. The second step is to generate a word sequence from such a task-independent semantic frame, using language-dependent generation rules to build coherent sentences.

In addition to the advantage of re-usability of the task-independent modules, some kind of discourse level processing can be put into the NLU and NLG modules. For example, focus shifting can be set up using syntactic information, which is always assimilated in the case of task-dependent interlingua for each language, and can be carried out in a task-independent way.

The disadvantage of the semantic frame conversion approach is that it increases the number of language-dependent modules. In the general interlingua approach (discussed in the previous section), the system developer must customise the language-dependent modules (NLU and NLG) for each task. However, in the semantic frame conversion approach two converters are required: one to transform task-independent frames into task-dependent frames (in the understanding side), and the other or to transform task-dependent frames into task-independent frames (in the generation side).

4.1.4 Dialogue-Control Centred Approach

If there is a pre-existing GUI-based Web application and the system developer wants to add a multilingual speech interface, a dialogue-control centred architecture can be a good solution. This architecture consists of a voice browser (which can be the compound of a language-dependent ASR, language-dependent simple NLU, template-based NLG, and TTS), a controller and a set of application control programs called *application logic*. Table 4.3 shows some possible choices for setting up this approach, while the overall architecture is shown in Figure 4.6.

The difference between the two former interlingua-based approaches and this one relies on a type of *control* information. In the interlingua approach, the content of the user's sentence is passed to an upper layer module which transforms it into an application command, and the contents of the system's response drives the NLG process. Thus using the interlingua approaches, the language-dependent representation must be assimilated at a certain module as *interlingua*.

On the contrary, in the dialogue-control centred approach, a control module termed *controller* in Figure 4.6, drives the whole dialogue process interacting with each module by a specific method. The controller calls an application program corresponding to the user's request and this program returns the result of the processing as a *status* to the controller. The contents from the back-end application are stored in specific objects that can be accessed by the response generation mechanism in order to fill template slots and generate

Table 4.3 Choices for the dialogue-control centred approach

	ASR	NLU	Interlingua	Context	NLG
Choice	Grammar-based LM	Simple Assignment	Not used	N.A.	Template-based
Language portability	Medium	Good	Excellent	–	Good
Task portability	Medium	Good	Excellent	–	Good

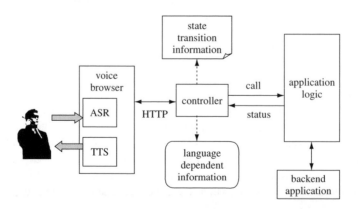

Figure 4.6 MLDS architecture based on dialogue-control centred application

sentences as system response. Therefore, unlike in the interlingua approach, language-dependent information can mostly be excluded and almost all the system modules can be language-independent developed.

The disadvantage of this approach is the inflexibility of the user's input and the system's behaviour. The ASR module is restricted since is implemented using a grammar-based approach with semantic interpretation tags embedded (e.g. SRGS shown in Section 4.1.1). In addition, the sentence generation is also restricted as is implemented using a template-based method. Moreover, it makes very difficult to deal with discourse phenomena such as ellipsis and anaphora. As a result, the dialogue management is generally based on a simple system-directed interaction strategy.

4.2 Multilingual Dialogue Systems Based on Interlingua

The essence of the interlingua approach in MLDSs is to design a semantic representation common to all the target languages. The basis comes from MT research, which implements modules to convert from natural language (e.g. English, Japanese, Spanish, etc.) to interlingua, and vice versa. For setting up these modules, system developers do not need to know foreign languages but only their native language and the specification of the interlingua, i.e. how to represent a semantic meaning in the interlingua. This approach is a promising solution especially if the number of target languages is large. For example, if the number is N, it requires 2*N modules to translate among the N languages. In comparison, the transfer approach, which consists of conversion rules between structural representations (e.g. parse trees) for each pair of languages, needs N*(N-1) rules. Despite the possible advantages of the interlingua, only a few MT systems have been implemented based on this approach. The reason is that it can be very difficult to establish language-independent semantic representations assimilating the language differences. For example, the range of target concepts may be different for the nouns, and there may be differences in tense and modality.

Despite these difficulties, some MLDSs use interlingua as the internal meaning representation, typically for the NLP output and the dialogue manager input, and for the output of the later and the NLG input. The architecture of an interlingua-based MLDS follows that of an ordinary SDS, in which the behaviour of each module is decided by the user's input and the system's state. This approach is appropriate for setting up the dialogue initiative (either user-initiative, system-initiative or mixed) for a variety of tasks (e.g. information seeking, planning, negotiation, etc.). Basically it needs language-dependent modules (for NLP and NLG) for each language.

In this section, we discuss some implemented MLDSs based on interlingua to show the main pros and cons of using this approach.

4.2.1 MIT Voyager System

Voyager is a MLDS implemented to provide regional information about Cambridge, Massachusetts (USA) (Glass et al. 1995). The system components for ASR, NLU, dialogue management and NLG, were developed at the MIT (Massachusetts Institute of Technology, USA) by the SLS (Spoken Language Systems) research group. For TTS, appropriate systems were used for each language. The overall architecture of the system is illustrated in Figure 4.7.

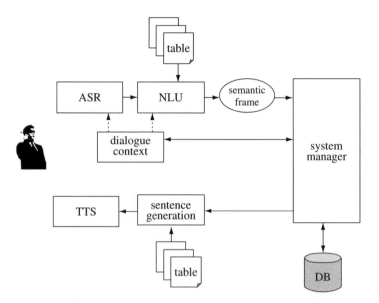

Figure 4.7 Architecture of the MIT Voyager system

The system provides information about distances, travel times, and directions between restaurants, hotels, banks, libraries, etc. as well as their corresponding addresses and telephone numbers. The interaction languages are English, Japanese and Italian.

The system was developed under the assumption that it was possible to obtain a language-independent semantic representation (interlingua) from task-oriented sentences, provided that the details of the sentences should not be represented, but rather the data in them required to construct application commands.

The language-dependent information is assimilated by the intermediate language (semantic frames), designed to avoid language-specific features. The dialogue management is performed using this intermediate language, which consists of "clauses", "topics" and "predicates".

4.2.1.1 Speech Recognition

The system uses the SUMMIT ASR system, which generates word hypotheses using an acoustic-phonetic network, phonological rules and a language model (Section 2.1.1.7). This model is a class-bigram for English and Italian, while for Japanese it is a category bigram in which the classes are defined by morphological categories. The acoustic-phonetic network is a context-independent mixture of up to 16 diagonal Gaussians (Section 2.1.1.6). The phonological rules are used to create a pronunciation network. The recognition is performed using an A* search (Rabiner and Juang 1993) that compares the acoustic-phonetic network against the pronunciation network and produces as output either an N-best sentence (Section 2.1.1.2) or a word-graph.

Setting up the ASR for several languages required using training corpora built for the target languages. The phonetic models were iteratively trained using segmental K-means, starting from the seed models (Rabiner et al. 1996). Although the English seed models were obtained by the manually-aligned phonetic transcriptions of the corpus, the models for the other languages (Japanese and Italian) were obtained seeding the phonetic models from their most phonetically similar English counterparts.

4.2.1.2 Probabilistic Sentence Understanding

For setting up the NLG, the system uses the TINA analyser (Seneff 1989; Seneff 1992a; Seneff 1992b; Seneff et al. 1992), which is based on a probabilistic parsing of the input (N-best sentence hypotheses produced by the ASR module). The result is language-independent meaning representation.

TINA performs a top-down rule-driven sentence parsing and trace mechanism to handle movement phenomena. The hand-written context-free grammar rules are converted to a set of transition probabilities which are acquired from the parse tree obtained from the training corpus. Feature parameters associated with the category names are used to construct the semantic representations.

In order to set up this language understanding component for a multi-language setting, it require syntactic rules for each language and a training corpus to estimate parameters for each rule. In addition, in order to ensure that the same semantic representation was obtained from sentences with the same meaning that were uttered in different languages, the rules were carefully constructed one by one.

4.2.1.3 Meaning Representation

In this systems, the language-dependent information is assimilated by the intermediate language (intertingua) which was implemented by means of semantic frames that simply consists of a clause, a predicate and a topic. The clause corresponds to a sentence unit which represents a dialogue act specific to the target task. Predicates correspond to verbs, adjectives, prepositions and modifier nouns, which represent content words. A topic corresponds to a content word which is part of the semantic frame obtained from the sentence, and is inferred from a hierarchical structure. For illustration purposes Figure 4.8 shows the semantic frame obtained for the sentence 'Where is the library near central square?'.

Dialogue management is carried out using this intermediate language, which is designed to avoid language-specific features.

The discourse-level processing is in general difficult using this approach. However, the system used a simple anaphora resolution mechanism that worked well for the location

```
Clause: LOCATE
   Topic: BUILDING
      Quantifier: DEF
      Name: library
         Predicate: near
      Topic: square
         Name: central
```

Figure 4.8   Semantic frame obtained by the Voyager system

guidance domain. This mechanism has two slots, the former indicates the current location of the user, and the second represents the most recently referenced set of objects. Such a mechanism allows the system to achieve the correct meaning from sentences such as 'What is their address?' and 'How far is it from here?'. However, a drawback is that the procedure to extract the present user's location is language-dependent and there are some cases where the most recently referenced objects cannot be assigned to the reference expression.

4.2.1.4 Language Generation

The sentence generation component of the Voyager system, called GENESIS, translates the language-independent semantic representation provided by the dialogue manager to a surface word sequence for each language using the lexicon, a set of sentence templates and a set of rewriting rules, which are language-dependent. All these language resources are stored in external files, which ease the addition of new languages. The lexicon represents a deep linguistic knowledge for each language at the word level. Each lexicon entry contains a part of speech (POS) (e.g. noun or regular verb), a stem, and various derived forms (e.g. the word "which" has several surface realisations in French depending on the gender and number). The sentence templates, which correspond to the topics, predicates and clauses of the semantic frames, are used recursively in order to construct the sentences for the system output. The rewriting rules are applied to carry out surface phonotactic constrains and contractions, such as selecting among 'a' or 'an' in English, reformulating 'a other' into 'another' in English, or 'de le' as 'du' in French.

4.2.2 MIT Jupiter System

The MIT SLS group has implemented several SDSs using the Galaxy architecture which is described below (Seneff et al. 1998). One of these systems is Jupiter, which was designed to provide weather information in several languages, e.g. English, German, Japanese, Mandarin Chinese and Spanish (Nakano et al. 2001, Zue et al. 2000).

4.2.2.1 Galaxy Architecture

Galaxy is the reference architecture of the DARPA Communicator Program. It is a kind of distributed client-server architecture in which each component is a server that performs a specific task within the spoken dialogue processing, such as speech recognition or language understanding. These servers communicate with each other via a central hub, which behaves as a router (Figure 4.9). Thus, this architecture is referred to as *hub-and-spoke* architecture.

The communication between the servers is achieved by exchanging frame-like representations. Each server is implemented in a *plug-and-play* fashion, following a common interface specification. The audio server captures the user's speech via telephone and sends the waveforms to the language-dependent speech recognition module, which produces a word graph and passes it to the frame construction module. The later carries out the language understanding and generates a language-independent.

E-form (i.e. meaning representation) as output in which the meaning representation is somewhat task-oriented. The context-tracking module resolves discourse-level problems.

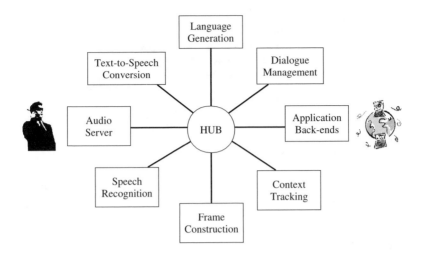

Figure 4.9   The Galaxy architecture

The dialogue management module, which is implemented as language-independent, receives the E-form and generates an E-form response.

Such hub-and-spoke architecture is especially useful for high-level SDSs in which several modules (e.g. the Belief, Desire and Intention (BDI) model) cooperate with each other, in contrast to simple task-oriented dialogue systems. In this architecture the information flow is controlled by an scripting language in the hub, allowing for the cooperation of components and the setting up of MLDSs. As we already know, an important aspect in these systems is the flow of content information and where to draw the line between the language dependent and independent components. Thus, from the viewpoint of information contents, the essence of the Galaxy architecture is the same as the interlingua approach described previously.

4.2.2.2 Language Understanding

As commented above, the frame construction module in the Galaxy architecture takes the word lattice provided by the speech recogniser and generates a language-independent meaning representation (interlingua) as output. A difference between the Galaxy-based systems and the Voyager system described above is in the task dependency of the meaning representation, since in Jupiter it is rather task-oriented. For example, Figure 4.10 shows the interlingua obtained by Jupiter for the sentence 'What is the weather in New York tomorrow?'.

Due to this task dependency, the meaning representation can easily be translated into an application command. Also, adding a new task or domain to the system requires little effort compared to the Voyager system. However, it requires a task-dependent language understanding module which is hardly reusable in another task or domain, given that the task-dependent meaning representation contains task-dependent concepts such as weather, date or city (shown in Figure 4.10). A different task may require a different representation of such concepts, either fine-grained (e.g. 'minute' in the *time* concept of a scheduling domain) or coarse-grained ('a.m.', 'p.m.' in the *time* concept of a weather forecast domain).

```
{c wh_query
     :topic {q weather
                    :quantifier "which_def"
                    :pred {p month_date
                            :topic {q date
                                    :name "tomorrow" }}
                    :pred {p in
                            :topic {q city
                                    :name "new york city"}}}
```

Figure 4.10   Interlingua used in the Jupiter system

Therefore, language understanding modules with a certain amount of complex rules are required for each language.

4.2.2.3 Towards Fluent Language Generation

In Jupiter, the GENESIS-II system (Baptist and Seneff 2000) was used for language generation. This system enhances the GENESIS system, described above for the Voyager system, by giving better control over the ordering of the sentence constituents and offering the capability of selecting context-dependent lexical realisation. The re-ordering of the sentence constituents is needed because for some languages, a constituent that is 'deep' in the semantic representation may be needed to appear at the beginning of the surface realisation (sentence). To address this problem, the system uses *pull* and *push* commands. The former allows to extract constituents from *deeper* in the semantic frames, while the second allows to defer certain substrings to a later stage of the surface realisation.

The lexical realisation takes into account that a word in one language may have several translations in other languages that are context-dependent. To address this problem, the system uses features that represent semantic context. For example, the English word 'near' can be realised in three different ways in Japanese for the weather domain. Although the default realisation is '*kurai*', if it is used to refer to climate records, then '*hotondo*' must be used, while if it is used in a expression containing a percentage, '*yaku*' must be used. This one-to-many realisation can be represented as follows:

```
near 'kurai' $:record 'hotondo' $:percentage 'yaku'
```

This method facilitates resolving the semantic ambiguities that can arise when generating sentences in different languages. Also, it can be useful to deal with the prosodic selection, by placing the $slow or $high features where appropriate.

4.2.3 KIT System

At the Pattern Information Processing Laboratory of the Kyoto Institute of Technology, Xu et al. (2003a, 2003b) proposed a MLDS architecture based on a case-frame conversion (Figure 4.11), which relies on the semantic frame conversion approach discussed in Section 4.1.3. The authors implemented Chinese and Japanese SDSs for multiple domains based on this architecture. Compared to the interlingua approach described above, the

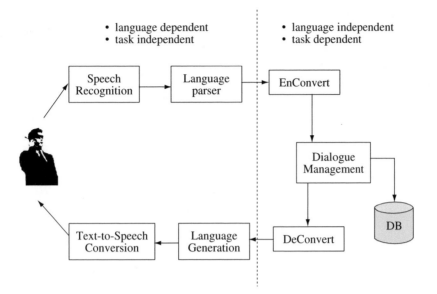

Figure 4.11 Architecture of the KIT system

case-frame conversion distinguishes task-independent language analysis and task-dependent command for frame construction. When using task-oriented meaning representations, such as E-form interlingua, it is difficult to ensure that compatible sentences for each language that have the same meaning are converted to the same semantic representation. This happens because the phrase structure and the location of modifiers differ in different languages. Therefore, the knowledge to extract the meaning of the same concept might be distributed as separate rules in the linguistic knowledge representation, e.g. using a semantic grammar.

In order to overcome this problem, the architecture of the KIT system separates task-independent linguistic knowledge, such as grammar rules and case frame construction rules, from task-dependent knowledge, such as case frame conversion rules. The case frame conversion rules describe the correspondence between the syntactic case (such as 'subject', 'object' or 'source') and the task-oriented case (such as 'date', 'place' or 'number'), which enables the reuse of linguistic knowledge in other tasks.

4.2.3.1 Speech Understanding

Figure 4.12 shows the result of the syntactic analysis of a Chinese sentence taken from a dialogue with the KIT system. The second line of the Figure contains a semantic representation that is described by a list of four terms:

- A word is used to reflect whether the sentence is an affirmative answer ('[Yes/Ok]'), a negative answer ('[No]'), an ambiguous reply ('[uhm]'), or an empty item ('[non]').
- A main verb.
- Modality information specifying the tense and the aspect properties of the sentence.
- A set of case elements (*slots*). Each slot indicates a relation between the main verb and the noun phrase/word in the slot, such as 'agent', 'object', etc.

```
Sentence:我需要景点介绍 (I need the sightseeing guide)
SS([[non],verb(需要)],[state],[agent([person:我]),object([function:景点介绍])]])
 ├─ s([verb(需要)],[state],[agent([person:我]),object([function:景点介绍])])
 │   ├── case(agent([person:我]))
 │   │    └──── nounPhrase([person:我])
 │   │            └──── pronoun([person],[pronoun(我)])
 │   │                    └──── 我
 │   ├── predicate([verb(需要)],[state])
 │   │    └──── verb(verb(需要))
 │   │            └──── 需要
 │   └── case(object([function:景点介绍]))
 │        └──── nounPhrase([function:景点介绍])
 │                └──── noun([function],[noun(景点介绍)])
 │                        └──── 景点介绍
 └─ period([endSent])
```

Figure 4.12   Example of task-independent semantic representation in the KIT system

For each input sentence, the KIT system conveys a list containing the above four terms from the speech interface to the dialog controller.

As semantic markers, this representation uses 'person' and 'function'. All noun phrases/words included in a sentence are assigned to some slots of the case frame considering the semantic markers of the noun phrases/words.

In general, the model of a case frame is represented as (V, M, C) where V denotes the main verb, M denotes the modality information that contains one or several modality elements, and C denotes one or several case elements, i.e.,

$$C = case\_id_1(case_1), \ldots, case\_id_n(case_n)$$

This model also can be written as $C = [C_1, \ldots, C_n]$.

4.2.3.2 Case-Frame Conversion Method

The case frame converter translates a task-independent and language-dependent semantic representation into a task-dependent semantic representation of the target language. The first step is called *case constituent ordering/pruning process based on a set of rules*, and the second *noun word replacing process based on a translation table*. Figure 4.13 shows a sample conversion of a Chinese case frame into a task dependent representation.

Using this method, the same semantic meaning can be represented in different case frames in different languages, e.g. using different case constituent or case element ordering. The case constituent ordering/pruning process deals with this problem. For example, in Figure 4.13 the case elements of the Chinese sentence (equivalent to 'I want to visit gardens') are different from those of the Japanese sentence, although they are intentionally equivalent. Thus, the converter uses a set of language-dependent, manually predefined rules to perform the conversion of main verbs, modality information and case constituents, and re-order the result.

All the words that the KIT system can accept are organised in a thesaurus. So, for all nouns that can fill in the slots of a case frame, it is easy to create word translation tables to map

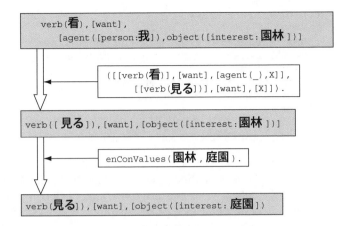

Figure 4.13 Example of case frame conversion

each word in a language into the appropriate word in another language. The word replacing process replaces nouns searching in the translation table to change a task-independent case frame into a task-dependent one before conveying it to the dialogue manager, and make an inverse conversion before the arguments are conveyed to the NLG module.

4.2.3.3 Frame-Based Dialogue Management

The KIT system has a set of 'topic frames' as a knowledge base about topics. A topic frame contains mutually related topics that might appear in a task domain. For example, in the sightseeing task, the name of a hotel, the room charge, the location and so forth constitute a hotel frame. For illustration purposes, Figures 4.14 and 4.15 show the data structure of a topic frame in C programming language. As can be seen in Figure 4.15, a topic frame consists of a name, a status id, a frame id, and a pointer to the data structure shown in Figure 4.14, which consists of a slot name, a value field (usually empty), a semantic marker, and a 'pre-command' and a 'post-command' that are triggered before and after the slot is filled, respectively. The priority item determines the priority of the slot.

The 'value' field is filled by a word, a numerical value or a pointer to other topic frame (as described in the 'ValueType' field). When the 'value' field is filled with a word, the semantic marker of the word is stored in the 'ValueType' field. The current version of the

```
typedef struct slot_message
{
  char *SlotName;    // slot name
  char *Value;       // value to fill in
  char *ValueType;   // semantic marker
  char *PreCommand;
  char *PostCommand;
  int  Priority;
  struct slot_message *prior, *next;
} SlotMessage;
```

Figure 4.14 Data structure of a slot in a topic frame

```
typedef struct Frame
{
  char *FrameName; // name of the topic frame
  char *status;    // suspend, closed, ongoing
  int FrameId;
  SlotMessage *SlotMessagePointer;
  struct Frame *prior, *next;
} FRAME;
```

Figure 4.15 Data structure of a topic frame

system uses four methods for filling an unfilled slot: (1) asking the user; (2) retrieving data from a database; (3) using the default value attached to the slot; (4) linking to another topic frame. The 'PreCommand' field stores the method actually used. When a slot has been filled, the dialogue controller makes one of the three following actions: (1) transfers the control to the slot with the next highest priority; (2) transfers the control to the linked topic frame; (3) provides the information requested by the user. The action to carry out is decided by the content of the 'PostCommand' field.

A task domain requires creating a set of task-dependent topic frames in advance. Since some slots are allowed to be filled with other topic frames, the set of topic frames forms implicitly a tree structure, which is called *static topic tree*. Throughout a dialogue, topics in a dialogue move within this structure as the dialogue proceeds. Thus, a dialogue forms a subtree of the static topic tree called *dynamic topic tree*, which represents the history of the dialogue about the topic.

4.2.3.4 Sentence Generation

For the sentence generation, the KIT system uses concatenated templates by applying rules recursively along with appropriate constraints (in terms of person, gender, number, etc.). For doing so, a set of templates for Chinese and Japanese were manually created. Each template consists of a sequence of word strings and/or slots to be filled with the data provided by the dialogue manager. For illustration purposes, Table 4.4 shows four categories of templates while Table 4.5 shows examples of templates for Chinese/Japanese. In Table 4.5, the variables 'x' and 'y' (in the text sentences) are replaced by the words in the input 'X' and 'Y', respectively, according to the conversion rules of the corresponding languages. Thus, 'X' and 'Y' represent words in the internal representation language, while 'x' and 'y' represent words in the target language.

4.2.3.5 Examples of Dialogue

The KIT system works in the domains of sightseeing recommendation, accommodation seeking and personal computer assembling guidance. The task structure of each domain is represented in topic frames. The task-independent parts of the system are common for all the tasks and domains. To conclude this section, Figures 4.16, 4.17 and 4.18 show, respectively, a Japanese accommodation-seeking dialogue, a Chinese recommendation sightseeing dialogue, and a Chinese PC assembling dialogue.

Table 4.4 Four categories of response template

Category	Meanings
Prompt	1. greetings to user. 2. prompt valid candidates to select. 3. prompt the selected information and the system dialog status.
Confirm	1. confirm current plan 2. confirm current dialog intention. 3. confirm a previous selected action.
Inquire	ask user to fill an unfilled topic frame slot.
Answer	answer to a user the result retrieved from a database.

Table 4.5 Examples of Chinese/Japanese templates

Template name	Input	Text sentence (for different languages)
promInit2PlsSelect	X	Chinese: 這是x 系統。y 中、汝需要那一項服務？ Japanese: こちらはxシステムです。yについて、何なりとお尋ね下さい。
promDayItinerary	X	Chinese: 現在讓我門来確定第x天的行程計画。 Japanese: x日目のコースを決めましょう。
promAlternative	X, Y	Chinese: x和y中、汝期望那一個？ Japanese: xとyと、どちらがよろしいですか？
inquireDays	non	Chinese: 汝打算化幾天時間? Japanese: 何日の御予定ですか？
inquireWhichIsSuit	X, Y	Chinese: 請問、汝喜歡x的y？ Japanese: xのyがよろしいですか？

```
S201: promInit2PlsSelect(京都観光案内,RetX)
      → こちらは、京都観光案内システムです。観光、交通、宿泊について、何なり
      とお尋ね下さい. (This is a guide system in Kyoto city, may I help
      you in sightseeingguide, traffic guide, or accommodation seeking?)
U201: 宿泊案内をお願いします。(I need the accommodation guide.)
      → [non], [普通動詞([願う])], [丁寧], [対格([宿泊案内])]
      → [伝達], [宿泊], [案内:[宿泊:[]]]
S202: promAlternative(ホテル,旅館,RetX)
      → ホテルと旅館と、どちらがよろしいですか？(Which do you like to stay
      at a hotel or a hostelry?)
U202: 旅館をお願いします. (I'd like to stay at a hostelry.)
      → [non], [普通動詞([願う])], [丁寧], [対格([旅館])]
      → [伝達], [宿泊希望], [宿泊:[宿泊希望:[旅館]]]
S203: inquireWhichIsSuit(どの辺り,旅館,RetX)
      → どの辺りの旅館がよろしいですか？(Which area do you like to stay in?)
U203: 洛北方面をお願いします。(I hope to find it in Rakuhoku-area.)
      → [non], [普通動詞([願う])], [丁寧], [対格([方面:[地域名:[洛北]])]
      → [伝達],[希望場所],[旅館プラン:[希望場所:[洛北]]]
```

Figure 4.16 Japanese recommendation sightseeing dialogue

```
S101: promInit2PlsSelect(京都観光案内,RetX)
      →這是京都市導遊系統。景点介紹、交通指南、住宿介紹中、汝需要那一項服務？
      (This is a guide system in Kyoto city, may I help you in
      sightseeing guide, traffic guide, or accommodation seeking?)
U101: 我需要景点介紹。(I need the sightseeing guide.)
      →[non],[普通動詞(需要)],[陳述],[動作主格([人称詞:我]),対格([案内名:景点介紹])]
      →[伝達]， [観光]， [案内:[観光:[]]]
S102: inquireDays(RetX)
      →汝打算化幾天時間？(How many days do you plan to spend?)
U102: 我打算化両天時間。(I will spend two days. )
      →[non],[普通動詞(化)],[希望],[動作主格([人称詞:我]),期間格([期間:両天時間])]
      →[伝達]， [期間]， [観光 [:期間 [:2]]]
S103: promDayItinerary(1,RetX),promAlternative(社寺 庭園,RetY)
      →現在讓我門来確定第一天的行程計画。寺廟和園林中、汝期望那一個？(Ok,let's plan
      the first day's itinerary. What are you interested intemples
      or gardens?)
U103: 我想看庭園。(I prefer to visit gardens.)
      →[non],[普通動詞(看)],[希望],[動作主格([人称詞:我]),対格([観光興味:庭園])]
      →[伝達],[観光興味],[行程:[観光興味:[庭園]]]
```

Figure 4.17   Chinese recommendation sightseeing dialogue

```
S301: promInit2Why(パソコン構築案内,RetX)
      →這是電脳組装指南系統 。請 告知 組装 電脳的用途 。(This is a guide system
      in PC-assembling. Please tell me the purpose to use a PC.)
U301: 我想做計算機通訊。(I want to do it for the Internet.)
      → [non],[普通動詞(做)],[希望],[動作主格([人称詞:我]),対格([利用目的:計算機通訊])]
      → [伝達]， [利用目的]， [構築:[利用目的:[パソコン通信]]]
S302: inquireMorePurpose(RetX)
      →還有其他用途嗎？(Do you use it for other purpose?)
U302: 我還想玩遊戲。(I want to use it to play game too.)
      → [non],[普通動詞(玩)],[希望],[動作主格([人称詞:我]),対格([利用目的:遊戲])]
      → [伝達]， [利用目的]， [構築:[利用目的:[ゲーム]]]
S303: inquireMorePurpose(RetX)
      →還有其他用途嗎？(Do you use it for other purpose?)
U303: 没有了。(No, No more.)
      → [没有了],[non],[non],[non]
      → [否定応答]， [non]， [non]
S304: promPurposeParts([パソコン通信,ゲーム],[ハード,メモリ ,ハード デスク,ディスプ
レイ],RetX),PromDecideParts(ハード,RetY),confirmDecidePart(ハードB,RetZ)
      →計算機通訊、遊戲所需的配件包括CPU、内存、硬盤、顕示 器。現在讓我門来選購CPU。請
      問、汝対C PU-B感覚如何 ？(For the purpose of the Internet and the game,
      The CPU,memory, harddisk and monitor are necessary. Now, lets decide a
      CPU, How do you think about CPU-B?)
U304: 価格是多少。(How much money is it?)
      → [non],[普通動詞(是)],[疑問],[主格([price:価格]),断定格([価格疑問:多少])]
      → [要求]， [値段]， [パーツ:[値段:[query]]]
S305: answerPrice(一万五千,RetX)
      →価格為一万五千日元。(It is fifteen thousand JPY.)
```

Figure 4.18   Chinese guidance for PC selection dialogue

4.3 Multilingual Dialogue Systems Based on Web Applications

Several research projects have examined the implementation of MLDSs (Section 4.2 commented some of them). In general, these systems use interlingua, i.e. a common semantic representation for the supported languages to assimilate multilinguality. Typically, these systems can deal with up to three languages for a single task. However, real-world multilingual voice services require a greater number of languages and a variety of tasks. For example, in order to offer an useful service for over 80% of the foreign visitors in Japan, a dialogue system should be able to handle around seven languages (Korean, four dialects of Chinese, English and Japanese). In addition, the system should provide services and information regarding the five most common tasks for visitors: shopping, restaurant recommendation, traditional culture, sightseeing, and enjoying nature (along with indispensable travel information, such as lodgings, transportation, and weather). Therefore, the implementation of multilingual and multi-task voice services is an important challenge for the research on MLDSs. However, it is difficult to construct such a multilingual and multi-task voice service by applying the existing methods, given that the interlingua approach requires at least understanding and generation modules for each language, which require language-dependent processes and knowledge.

In this section, we describe a method for internationalising voice Web systems based on RDF/OWL[6] and a Web application framework. In order to deal with multilingual and multitask problems, the internationalisation of the Web application framework separates language-independent information, i.e. the view component of the MVC (Model-View-Controller) model (to be studied in Section 4.3.2.1), from language-dependent information, such as templates to generate output sentences. However, two major problems remain with respect to this internationalisation: the generation of noun phrases for each language, which fill the empty slots of sentence templates, and the generation (or preparation) of speech grammars for the user input, which is not a problem in the case of GUI-based systems. In order to deal with these problems, the method we are discussing in this section uses an ontology of descriptions for the concepts and instances in the task domain using RDF/OWL.

In the Web application framework, it is important to follow standardised technologies for various aspects of the systems. Thus, we describe an implementation of MLDSs based on Web applications following speech technology standardisations and other Web-related standard technologies. As current GUI-based Web applications are developed paying attention mostly to the internationalisation of contents, for setting up MLDSs using the Web base technologies (e.g. Java, Servelet, etc.), a possible framework is Struts[7], which allows that output contents and control information can be written in a declarative form. Also, language-dependent descriptions are not mixed in the interaction with the back-end application.

In the following sections we first set the problems for practical deployment of MLDSs and analyse the requirements for the setting up. Then, we describe in detail how to meet these requirements using the Web application framework.

---

[6] RDF (Resource Description Framework) is a W3C standard for describing resources on the Web. OWL (Web Ontology Language) is an exact description of the information on the Web and the relationships between that information.

[7] http://struts.apache.org

4.3.1 Requirements for Practical Multilingual Dialogue Systems

As discussed at the beginning of this section, practical MLDSs should deal with several languages and a variety of tasks and domains. Based on these requirements, a question might be: to what extent should we require a great system's flexibility for all the languages and tasks?

On the one hand, in terms of languages, research SDSs can handle some kind of spontaneous speech in the input, including multiple information chunks, filled pauses, error correction, etc. In a mixed-initiative interaction strategy, in which the user and the system can take the initiative when needed, such spontaneous speech appears usually in the first sentence of the dialogue and when specific initiative-taking sentences are uttered. In order to deal with such a spontaneous speech, many SDSs use a statistical language model, which requires the analysis of a previously-collected dialogue corpus. However, in order to set up a MLDS, such corpus may be hard to obtain for all the target languages. In this case, grammar-based speech recognition can be a practical choice. Furthermore, in order to avoid recognition errors caused by spontaneous phenomena, the user's input should be restricted to a word or phrase, so that grammar rules can be easily written or generated from some kind of task knowledge representation. As a result of this input restriction, a system-directed interaction strategy with rule-based language modelling can be a realistic solution for implementing such a MLDS.

On the other hand, in terms of tasks, task-related components of SDSs tend to be tuned for the spoken interaction. If the new system is based on a pre-existing Web application, it can benefit from the pre-existing service logic of the application. However, in general it is a very time-consuming work to tune up many tasks already implemented with GUI-based Web applications for a new, spoken interaction. When considering the initial deployment of the system, it will probably start with two or three simple tasks that can be extended later. Thus, task extensibility (adding new tasks or modifying tasks), language extensibility, and easy modification of the dialogue flow are essential features to take into account in the design. The interlingua approach causes maintenance and task extension problems irrespective of the interlingua level, as discussed in Section 4.1.1.5. Therefore, a dialogue-control centred architecture (Section 4.1.4) can be a promising choice for the implementation.

From the viewpoint of technical needs, the progress in standardisation of speech-related technologies is utterly different from the early years of research on MLDSs. Nowadays, several companies (e.g. IBM, Microsoft, Nuance, etc.) provide speech recognisers and synthesisers for multiple languages, which basically follow, or are expected to follow in the near future, the (mainly W3C's) standardisations of grammar, semantic representation and synthesised speech control. Consequently, developers of MLDSs can design their systems presupposing the existence of speech recognition and synthesis modules for the target languages, with application interfaces following common internationalised representations.

4.3.2 Dialogue Systems Based on Web Applications

4.3.2.1 The MVC Model

The MVC (Model-View-Controller) model[8] (Figure 4.19) was originally developed for the design of graphical user interfaces, and has become the most popular model for Web

---

[8] http://www-128.ibm.com/developerworks/library/j-struts/?dwzone=java

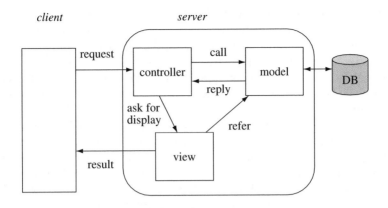

Figure 4.19 MVC model

applications. It separates the business logic (Model), which interacts with the back-end applications, the information representation (View), which is typically HTML in GUI-based Web applications, and the interaction flow management (Controller). JSP (Java Server Pages) can be used in the View component to be called from the Controller, allowing that the contents (data) of the applications' responses can be stored in objects that are accessible from the tag library in JSP. Moreover, JSP is not limited to generate HTML since can be used to generate any XML-based language. Therefore, by using this model, the Web application can be easily implemented and used by multiple devices, such as PCs (HTML), mobile telephones (WML[9]), and voice channels (VoiceXML[10] or SALT[11]).

The Model component is responsible for the business logic, which typically interacts with a database or a Web server outside the system. This component can be developed as a common program with a graphical interface and a voice interface.

The View component controls the interaction with the user. If the user's client is a GUI-based Web browser, then JSP can be used to generate HTML. Furthermore, since JSP can deal with any XML-based language as a generation target, it can also be used for setting up the voice interface creating contents (data) dynamically by calling the application logic.

The Controller component accepts information from the View component, requests processing from the appropriate Model component, receives the resulted status from the Model component, and calls for a suitable View component for the output. In order to handle the dialogue according to the result of the application access, the Controller uses a state transition rule which describes the current state (called by the input View component) and a conditional branch to the next state (by calling the output View component). The Controller's state transition rules are defined declaratively in a configuration file and therefore do not require any procedural description

Based on the MVC model, a number of Web application frameworks have been developed. Here, the concept of framework is opposite to the library approach typically used in

---

[9] http://www.wirelessdevnet.com/channels/wap/training/wml.html

[10] http://www.w3.org/TR/voicexml20/

[11] http://www.saltforum.org/

system development by means of C programming language. Ordinary library-based system development uses several types of libraries, such as mathematical calculations, string handling, list operations, and network communications. Web application programs can call such libraries in order to perform functions. However, in the MVC framework, the framework controls the task flow and calls programs written by the Web application developer. In other words, Web application developers only need to write small functions for each task, which are called by the framework.

Struts[12] is a popular Web application framework based on the MVC model. It provides a Controller component and describes how to connect the processing results of the Model component with the View component. In addition, it provides a data handling method which stores the input data from the View component, manipulates the data from the Model component and maps the resulted data to the View component for output. An outline of this processing is shown in Figure 4.20, which can be described as follows:

1. The user's client submits a request to the Controller which is called *action* servlet. A typical user's client request is a HTTP 'get' command.
2. The controller accepts the request and stores its contents (data) in an action form bean which is implemented in a Java class.
3. The controller calls the requested action class, which typically links to an external data resource such as a database or Web page.
4. The action class uses the action form bean in order to do the task.
5. The action class sets the result to the action form bean and returns the resulted status to the Controller (e.g. 'success' in case of appropriate search result, 'failure' if no data returned).
6. The controller decides which page should be sent to the client according to the resulted status, in accordance with the state transition rules stored in the action configuration file.

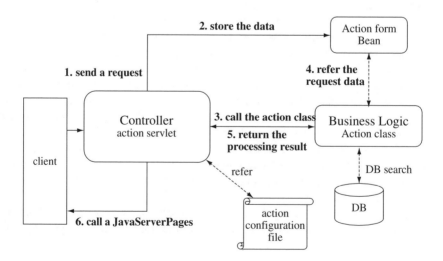

Figure 4.20 The Web application framework

---

[12] http://struts.apache.org

To implement the business logic in the action classes, the Web application developer must define the data structure of the target application in the action form bean, and design the view in the Java Server Pages. The overall control information must be defined in the action configuration file, in the form of *action name* called from the input View component, branch information based on the status returned from the application logic, and URI (Uniform Resource Identifier) of the next page to be passed to the View component.

4.3.3 Multilingual Dialogue Systems Based on the MVC Framework

Figure 4.21 shows a possible MLDS architecture based on the MVC framework. In the Model component, the contents are expressed in a specific language, and are retrieved according to the value of the action form bean. Therefore, the language used in the contents description and the value of the action form bean must be identical. In addition, there should be no language-dependent code (e.g. conditional branches with a test of equality of language-dependent strings) in the action class program. Such exclusion of language-dependent information is in common with GUI-based internationalised Web applications.

The View component contains two types of language-dependent information: for sentence understanding and generation, respectively. The first is discussed in this section while the second is addressed in Section 4.3.3.2. In the understanding process, the View component contains a grammar to handle the user input which is not used in GUI-based applications, and thus is obviously language-dependent. Generally speaking, writing comprehensive grammar rules that allow the user speak freely is very difficult for non-experts in linguistics, even for simple tasks. However, if the user input is restricted to a word or a phrase, such grammar rules can be easy to write, or can be semi-automatically generated by the domain ontology, given that the non-terminal symbols of the grammar rules can be mapped to the concepts of the task

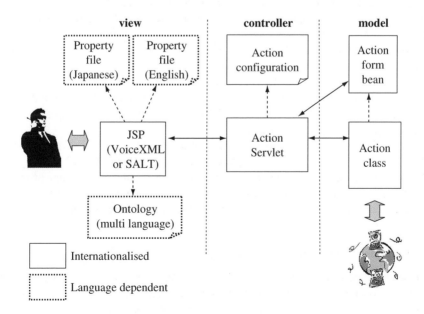

Figure 4.21 MLDS system architecture based on the MVC framework

```
<action-mapping>
      <action path="/Reservation"
                    type="ReservationAction"
                    name="ReservationForm"
                    scope="request">
            <forward name="success" path="/confirm.jsp" />
            <forward name="failure" path="/retry.jsp" />
      </action>
</action-mapping>
```

Figure 4.22   Example of Action Configuration File

domain. For example, by using the mixed-initiative interaction strategy in VoiceXML (to be studied Section 6.1.2.1 and Appendix A), the user can enter free combinations of values to a certain extent, which can be interpreted by such simple grammar rules. Thus, the system developer must prepare the vocabulary for each language maintained in the task ontology (described in the next section) and a small number of phrase rules for each language. This makes a grammar file that can be called by the JSP in the understanding process.

In the Controller component, the set of state transition rules in the configuration file is described using the status information passed by the action class, and as such, contains no language-dependent information.

4.3.3.1 Dialogue Management Using the Web Application Framework

Building a Web application with Struts requires creating an action configuration file, an action form bean and JSP. The action configuration file is used by the Controller and specifies the action mapping rules which connect the Universal Resource Identifier (URI) requested by the client with the name of the action class program. In addition, this file specifies the View page according to the returned status from the action class.

Figure 4.22 shows a sample action configuration file, which indicates that if the URI requested from the client is '/Reservation', then the 'ReservationForm.class' must be activated and the processing result (either 'success' or 'failure') returned as status. The returned status determines the next dialogue page (JSP).

An action form bean (Figure 4.23) is a Java class that preserves the value of the field variables. The value submitted from the user's client is stored in the bean and is used by the action class and View component (i.e. JSP). The action class is a simple class that has setter/getter methods for the field variables.

Java Server Pages can contain XML forms that may embed Java programs for communicating actions from beans or back-end applications, for example via VoiceXML (Figure 4.24) or SALT. The messages the user will receive can be written in external property files (Figure 4.25) prepared for each language. This way, by separating the View contents and the language-dependent messages, Web applications can be internationalised.

4.3.3.2 Response Generation by Combining Chunks

For the sentence generation, the View component can employ the traditional template and slot filling mechanism. There are three ways of adapting the templates' slots to be filled with content words under the multilingual framework (Table 4.6):

- Managed by the TTS module, using the same representation for all the target languages with SSML tags such as 'number', 'currency', 'date' and 'time'.
- Handled by the phonetic description (via IPA), in the case of proper nouns.
- Using the domain ontology, in the case of a noun phrase.

Using the same principle employed in example-based Machine Translation (MT), the meaning of the sentence can be considered consisting of chunks, which are minimal semantic units. Therefore, the sentence generation can be divided into two stages: chunk generation and chunk combination. The three types of multilingual adaptation methods mentioned above are used for the chunk generation process. For the chunk combination process, the patterns will surely differ for the different target languages. Therefore, the system developer must prepare these patterns in advance in addition to the sentence templates, in such a way that in the surface realisation, each chunk combination pattern can be filled similarly as the templates for the sentence generation.

```
public final class ReservationForm extends ActionForm {
        private Date checkIn;
        private Date checkOut;
        private String roomType;
        ...
        public void setCheckIn(Date checkIn) {
            this.checkIn = checkIn;
        }
        public Date getCheckIn() {
            return checkIn;
        }
        ...
}
```

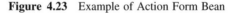

Figure 4.23   Example of Action Form Bean

```
<%@page contentType="text/xml; charset="Shift_JIS" %>
  <%@ taglib uri="/WEB-INF/struts-bean.tld" prefix="bean" %>
  <vxml version="2.0">
   <form>
    <field name="confirm" type="boolean">
     <prompt>
      <bean:message key="confirm.open"/>
     <bean:message key="confirm.checkIn"/>
      <bean:write name="ReservationForm" property="checkIn"/>,
     <bean:message key="confirm.checkOut"/>
     <bean:write name="ReservationForm" property="checkOut"/>
      ...
     </prompt>
    </field>
    <filled> ... </filled>
   </form>
  </vxml>
```

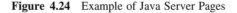

Figure 4.24   Example of Java Server Pages

```
English property file
confirm.open=Your reservation is as follows.
confirm.checkIn=Your check-in time is{0}.
...
Japanese property file
confirm.open=予約内容は以下の通りです
confirm.checkIn=チェックインのお時間は {0}
...
```

Figure 4.25   Example of Property files

Table 4.6   Multilingual treatment of contents words

Class	Multilinguality
Inside TTS	Using say-as element of SSML
Proper noun	Writing IPA character at phoneme element of SSML
General noun	Surface realisation of the relation between concepts on ontology

4.3.4 Implementation of Multilingual Voice Portals

Following the MVC framework and multilingual adaptations mentioned above, Omukai and Araki (2004) implemented a preliminary voice portal (kind of dialogue system with a web interface) for Japanese and English called *Kyoto Voice Portal* (Figure 4.26). The system was implemented in SALT, fully for Japanese and partially for English. Its task

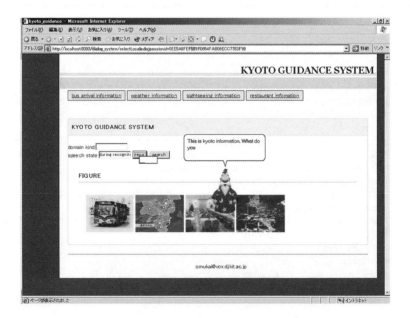

Figure 4.26   Screenshot of the Kyoto Voice Portal

was to provide weather information, bus arrival guidance and restaurant recommendation in Kyoto area, Japan. At the beginning of the dialogue, the user specifies the intended task by a menu selection dialogue and continues to perform each task, using a system-directed interaction strategy. The back-end application is a Web site that provides the three service types mentioned above.

For grammar maintenance and noun phrase generation, the systems uses an ontology that was prepared for each task containing relationships between concepts. For example, in the weather information task (Figure 4.27), the concepts *prefecture* and *area* have a 'sub-component' relationship. This relationship has as properties a Japanese and an English noun phrase templates, which are expressed as shown in Figure 4.28 specifying 'domain', 'range' and 'expression' elements. The 'expression' indicates the surface linguistic realisation for each language using the node id's for the 'domain' (#id0) and 'range' (#id1). Thus, the expression is '#id0 #id1' for Japanese and '#id1 of #id0' for English. This way, if the values for the *prefecture* and *area* fields in the action form bean were 'Kyoto' and 'southern part', respectively, the ontology could be used to generate the English noun phrase 'southern part of Kyoto'. The multilingual dictionary of the system was created using mappings of concepts

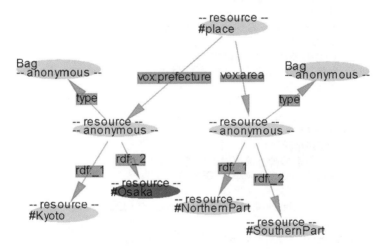

Figure 4.27 Part of the ontology of the weather domain

```
<rdf:Property rdf:about="#subclass">
  <rdfs:domain rdf:resource="#id0"/>
  <rdfs:range rdf:resource="#id1"/>
  <vox:expression xml:lang="ja">
    #id0 #id1
  </vox:expression>
  <vox:expression xml:lang="en">
    #id1 of #id0
  </vox:expression>
</rdf:Property>
```

Figure 4.28 Noun phrase rule representation

and containers of literals, as shown in Figure 4.29 for the concept '#kyoto' (for Japanese and English).

Following this method, bus arrival guidance and restaurant recommendation tasks were easily implemented. The method can ease the rapid development of MLDSs especially in

```
<rdf:Description rdf:ID="#kyoto">
  <vox:dictionary>
   <rdf:Alt>
    <rdf:li xml:lang="ja"> 京都 </rdf:li>
    <rdf:li xml:lang="en">Kyoto</rdf:li>
   <rdf:Alt>
  <vox:dictionary>
<rdf:Description>
```

Figure 4.29 Multilingual dictionary: Correspondences for 'Kyoto'

S1:こちらは京都案内システムです. 市バス接近情報,天気予報,
　　レストラン, 観光案内から知りたい情報をおっしゃってください.
U1:天気予報
S2:知りたい地域をおっしゃってください
U2:京都の南部
S3:いつの情報が知りたいですか
　　今日, 明日, 明後日からお選びください
U3:今日
S4:どの情報が知りたいですか
　　天気予報, 降水確率, 気温からお選びください
U4:降水確率
S5:京都の南部の
　　今日午前の降水確率は20%です
　　今日午後の降水確率は30%です

Figure 4.30 Dialogue example in Japanese

S1: This is Kyoto information System
    What do you want to know?
    bus, weather, restaurant, sightseeing?
U1: Weather.
S2: Where do you want to know?
U2: South of Kyoto.
S3: At when do you want to know?
    Today, tomorrow, the day after tomorrow?
U3: Today.
S4: What Information do you want to know?
    weather, chance of rain, temperature?
U4: chance of rain
S5: In south of Kyoto,
    today in the morning, the chance of rain is 20%.
    today in the afternoon, the chance of rain is 30%.

Figure 4.31 Dialogue example in English

the case of already existing GUI-based Web applications. To conclude this chapter, two sample dialogues for the weather domain are shown in Figures 4.30 (Japanese) and 4.31 (English).

4.4 Summary

In this chapter we discussed architecture and implementation issues of MLDSs. First, we addressed considerations in the setting up of these systems, focusing on speech recognition and synthesis, language understanding and generation, level of interlingua, and dialogue context dependency. Next, we focused on several approaches to the implementation, two of them based on the interlingua approach (general interlingua and semantic frame conversion) and the last one based on a dialogue-control centred approach. Secondly, we surveyed some systems based on interlingua (Voyager, Jupiter and KIT), especially focusing on the set up of the language independency in the semantic representations, the management of the differences in each language, and the generation of multilingual sentences. Thirdly, we explained the advantages and disadvantages of the system architecture based on the Web application framework compared to the interlingua architecture, and described how to construct MLDSs following the internationalisation methodology, which is widely used in GUI-based Web applications. We addressed requirements for practical MLDSs, presented the MVC framework and described systems based on this approach paying special attention to the dialogue management and the response generation. The chapter ended by addressing the implementation of multilingual voice portals, focusing as sample on the *Kyoto Voice Portal*.

4.5 Further Reading

Web Application Framework

- Dudney (2003)

Semantic Web

- Antoniou (2004)

VoiceXML

- Larson (2002)

5

Dialogue Annotation, Modelling and Management

5.1 Dialogue Annotation

The annotation (also called *tagging* or *labelling*) of dialogue corpora aims at analysing the features of real dialogues between people in determined tasks in order to build computational models that carry out these tasks automatically by dialogue systems. During the annotation it is usual to assign one or more tags to each interaction made by the participants in the dialogue.

5.1.1 Annotation of Spoken Dialogue Corpora

The annotation of spoken dialogue corpora can be carried out at the morphosyntactic, syntactic, prosodic or pragmatic levels (Gibbon et al. 2000). The *morphosyntactic* annotation, also known as word-class, POS (Part-of-speech) or grammatical annotation, aims at assigning to each word in the corpus a tag representing the category of words it belongs to. These tags can be organised according to a type hierarchy with attributes and values, such as noun, pronoun, adjective, verb, adverb, interjection, etc. Thus, the tag 'name' can have the attributes and values *Type* (common or proper), *Number* (singular or plural), *Gender* (male or female), etc. The set of tags and the guidelines to apply them are called a *tagging scheme*. Several schemes have been developed to carry out this type of tagging; for example, Burnard (1995) proposed one based on SGML tags, which was used to annotate the British National Corpus. One drawback of this tagging scheme is that it was developed to annotate written texts, which do not contain phenomena typical in the spoken spontaneous expression, such as lack of fluency (e.g. '*hum* . . .'), interruptions, false starts (e.g. '*On Mond* . . . *Tuesday*'), so that it is not very suitable to annotate spoken dialogues.

The *syntactic* annotation aims at linking each sentence in the corpus to a hierarchical structure representing the relationships between its words. The main problems with this type of tagging are also concerned with the fact that sentences spoken spontaneously can be ungrammatical because of incomplete segments (e.g. '*She was going into the* . . .'),

Spoken, Multilingual and Multimodal Dialogue Systems: Development and Assessment   Ramón López-Cózar Delgado
and Masahiro Araki   © 2005 John Wiley & Sons, Ltd

word repetitions (e.g. *'Oh ... I don't think ... I don't think ... ever want to see mine'*), unintelligible fragments (e.g. *'No but ... twenty-one, aren't you?'*), etc.

The *prosodic* annotation aims at dividing the speech into units in order to obtain information about the pronunciation of words, pauses, rhythm, etc. It is usual to carry out this type of tagging using graphic tools which allow observation of the speech signal. A typical tagging scheme at this level is ToBI (Tones and Break Indices),[1] which is the 'standard' scheme for tagging utterances spoken in American English, although it has also been applied to other varieties of English and other languages. One advantage of using this scheme is the large number of available training material and tools. Using the annotation tool, the tagging is carried out observing the speech signal, the fundamental frequency (F0) and tags grouped into four categories: tones, orthographic transcriptions, break indexes and events not related with tones (e.g. extralinguistic or paralinguistic, such as the quality of the speech signal). Tones may be high (H) or low (L) and may be either part of an accent or be signalling a boundary. Accents may contain one or more tones. The standard ToBI scheme for American English (called E_ToBI) contains five pitch accents: H* for peak accent, L* for low accent, L + H* for scooped accent, L* + H for rising peak accent, and H+!H* for clear step down onto the accented syllable.

The *pragmatic* annotation aims at obtaining an abstract description of the intention of the person taking part in the dialogue. Several tagging schemes have been proposed, mostly based on *speech acts* theories, which define the meaning of sentences considering that they are constituted of small functional fragments, for example, 'greeting', 'request', 'suggest', 'accept', 'negate', 'confirm', etc. (Searle 1975). Typically, speech acts are grouped into five categories:

- *Directive*. The speaker wants the dialogue partner to carry out some action.
- *Compromise*. The speaker wants the dialogue partner to carry out some action in the future.
- *Expressive*. The speaker expresses his feelings about something.
- *Representative*. The speaker expresses his beliefs about something.
- *Declarative*. The speaker provokes a change in the situation.

In this type of tagging, dialogues are divided into turns. There are several conceptions about what we must understand as a *turn* of a dialogue. For simplicity, we will consider it is the set of sentences uttered by a speaker until another dialogue partner starts to speak.[2] According to classical speech acts theories, each dialogue is labelled with just one tag. Obviously, this represents a problem for the tagging of SDSs since using just one turn, the speaker may 'respond', 'accept' and 'promise', for example. Several tagging schemes have been proposed to resolve this drawback. One of them is called DAMSL[3] (Dialogue Act Markup in Several Layers), developed by DRI (Discourse Research Initiative). Using this scheme, each sentence is annotated at four different layers:

- *Communicative status*. This layer contains tags indicating whether the sentence is understandable and has been completely uttered (e.g. 'not interpretable', 'abandoned', etc.)

---

[1] http://www.ling.ohio-state.edu/~tobi/

[2] For more detailed treatment of a turn, slash-unit was proposed as a unit of pragmatic annotation (Meteer 1995; http://www.ldc.upenn.edu/Catalog/docs/treebank3/DFLGUIDE.PDF).

[3] http://www.cs.rochester.edu/research/cisd/resources/damsl/RevisedManual/RevisedManual.html

- *Information level and status.* This layer contains tags concerned with the semantic content of the sentence and its relation to the task in hand, as well as tags concerned with the communication management. For example, there are tags to annotate sentences used to initiate the conversation (e.g. 'Hello'), finish it (e.g. 'Goodbye'), keep the contact (e.g. 'Can you hear me?').
- *Forward-looking function.* This layer contains tags to capture the effect that a sentence produces in the subsequent dialogue. For example, with a sentence the speaker may request information, ask a question, etc.
- *Backward-looking function.* This layer contains tags to capture how a sentence relates to the previous dialogue. For instance, the sentence can be used to respond, accept, deny, etc.

Another pragmatic annotation scheme has been proposed by the Japanese Discourse Research Initiative (JDRI 2000). It relies on dialogue act annotation and uses basic patterns to exchange dialogue segments, such as Initiation + Response + Follow-up. The subcategories of each exchange element are dialogue act tags (e.g. subcategories of the 'Initiation' exchange segment are 'Request', 'Suggest', 'Persuade', 'Propose', etc.). In this scheme, the multidimensional phenomenon of dialogue acts addressed by DAMSL is limited to a combination of Response/Initiation type tags, which simplifies the tag structure. Another characteristic of this scheme is that the dialogue segment tags consist of a Topic Break Index (TBI), a topic name and a segment relation. The TBI represents the degree of topic dissimilarity between the dialogue segments, which is determined by the sequence of dialogue act tags patterns. A TBI can take the value either 1 or 2 (value 2 means the boundary is less continuous in terms of topic). The *topic name* is labelled by the coders' subjective judgement. Finally, the *segment relation* indicates the relation between the preceding and the following segment, which is classified as 'clarification', 'interruption' or 'return'.

Several tools can be found in the literature to carry out the dialogue annotation, for example, DAT (Dialogue Annotation Tool).[4] As can be observed in Figure 5.1(a), the tool shows on one window the data about the annotating process (e.g. status, annotator, annotation date, problems, scenario, etc.). Figure 5.1(b) shows another window of the tool in which it shows the dialogue turns: when the annotator clicks on one of them, all the applicable tags are also shown, thus facilitating the annotation process. The tool can be downloaded from the Internet[5] including two sample dialogues, one from TRIPS (the Rochester Interactive Planning System) and the other from the Verbmobil dialogue corpus.

Other annotation tools available on the Internet are Alembic, AnnoTag, Carletta Python Coders, DiET, EMU, ESPS, Fringe, LTG XML Tools and NB (information about these tools is available at the MATE[6] Deliverable D3.1 Specification of Coding Workbench).

Once the dialogue corpus has been tagged, the dialogue system designer can use the diverse information stored in it to create the computational model to implement the system. Among other information, the analysis of the tagged corpus provides information about the vocabulary, syntactic and semantic structures typically used in the domain, the dialogue strategies used by the human service provider to interact with the user, and the types of

---

[4] http://www.cogsci.ed.ac.uk/~amyi/mate/dat.html
[5] http://www.cs.rochester.edu/research/cisd/resources/damsl/
[6] http://www.cogsci.ed.ac.uk/~amyi/mate/report.html

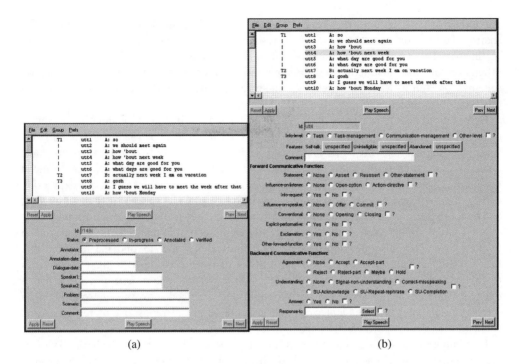

(a)   (b)

Figure 5.1   DAT annotation tool: (a) Dialogue annotation data. (b) Turns and applicable tags (reproduced by kindly courtesy of Dr Core, University of Southern California, Institute for Creative Technologies, Marina del Rey, CA, USA)

sentence the system must be able to generate to interact with the user. Therefore, all this information is very useful in the setting up of the ASR, NLP, the dialogue manager, and the NLG modules of the system. The dialogue manager must be designed with the most appropriate interaction and confirmation strategies, as well as whatever fallback strategies needed to repair interaction errors. It is important to note that according to several studies, people behave differently when interacting with a machine, so that the obtained model is normally refined to consider the features of a real human-to-machine interaction. The WOz technique (described in Section 3.2.1.1) is typically used for this purpose. The refined model is then used to drive the system-user interaction.

5.1.2 Annotation of Multimodal Dialogue Corpora

At the moment, the annotation of multimodal dialogue corpora seems to show a trend in research groups to create their own corpora, codification and annotation schemes. Mostly, these resources are created *ad hoc* without following established guidelines, are usually very domain-dependent, and are created for specific purposes without the initial aim of sharing with other groups. They are kept in private databases or, if exposed on Web pages, access is restricted or not publicly announced. Since multimodal annotation is a very costly task, this lack of availability is a waste of time and effort as it implies doing again the same work previously done. Therefore, it is very important to stimulate the development of means for sharing the resources created by different researchers so that they can easily be reutilised. In the past few years several initiatives have been carried out with the aim to provide efficient means for constructing and sharing multimodal corpora, codification and annotation schemes, as well as to establish easy to understand, general guidelines for the creation of this type of resources.

One of these initiatives comes from the European Natural Interactivity and Multimodality (NIMM) Working Group of the joint EU-HLT/US-NSF project International Standards for Language Engineering (ISLE) (Dybkjær and Bernsen 2002). The NIMM participants made a survey of existing resources in the field of multimodal corpora which includes 36 facial resources, 28 gesture resources, 7 descriptions of annotation schemes for facial expression and speech, 14 descriptions of annotation schemes for gesture and speech, 12 annotation tools and ongoing projects for tool development. Furthermore, they provide means to communicate new resources via the Web so that even recently created corpora and schemes can be added to their lists. They also developed some guidelines about how to find the resource that best fits the particular needs of a new application, of those considered in their repositories (Dybkjær and Bernsen 2004). The goal of the ISLE NIMM reports is to provide a list of existing resources while standards are proposed and accepted, covering the gap between a past of isolated task-dependent work and a future of global repository of resources and standard tools for corpora creation and annotation.

It is important to differentiate between *coding scheme* and *annotation scheme*. The former describes how to encode the studied phenomena. For example, the basis of gesture coding schemes is the study of the articulations' movements. The coding is usually carried out distinguishing several phases (e.g. initiation, expressive and retraction) which receive different notations in literature. Many coding schemes can be found in the literature, but to fulfil the growing need for globally understandable and standard schemes, they must not be constructed tied to specific tasks but be open to totally different usages. According to Wittenburg et al. (2002a), the unique way of reaching these requirements is by means of an iterative coding approach, where the scheme can be refined as innovative usages in new domains appear.

On the other hand, *annotation schemes* describe the semantics in structural terms, i.e. the role of annotation is to give a linguistic interpretation to each structural aspect. For example, in terms of gesture annotation, the annotation scheme should provide an interpretation for each gesture. This is not a trivial task as each gesture can have different linguistic interpretations, making the annotation complexity very high. For example, Martell et al. (2002) reported an annotation rate of 3 seconds per hour using their FORM system for gesture annotation. This system uses a series of tracks to represent different structural aspects such as location, shape, orientation and movement. It uses different levels of abstraction to group tracks of the different parts of the body together and classify them into subgroups. Such a classification allows users to isolate an anatomical movement from the global gesture.

The Abstract Corpus Model (ACM) described in Wittenburg et al. (2002a) is an object-based corpus model which permits a remote and simultaneous access to objects and classes via RMI (Java Remote Method Invocation), allowing a collaborative annotation and classification of annotations into classes. The annotation classes, called *tiers*, describe the same phenomenon, share the same metadata values and can be customised to the users' requirements (the users can define their own *tiers*). The metadata values can be used to include the annotated corpus in corpus browsers like the one proposed by IMDI[7] (EAGLES/ISLE Meta Data Initiative). The ACM model can be exploited by a Java and XML tool called EUDICO[8].

5.1.2.1 Annotation of Speech, Gestures and Other Interaction Modalities

Mostly multimodal dialogue corpora are comprised of gesture and speech. This is why many studies have been carried out into integrating both modalities into the same annotation and coding scheme. The availability of public annotated gesture and speech corpora not only has the advantage of easy sharing of resources (as previously commented), but also the possibility of studying the effect of non-linguistic cues (hand movement in this case) in supporting and completing the information expressed verbally. This becomes very important and demanding when applied to sign languages where gestures have their own morphology, syntax, semantic and even some kind of 'prosody' (Wittenburg et al. 2002a).

In addition to annotation and coding schemes for gesture and speech, it is also very important to use some general guidelines for gesture extraction and recognition. Kettebekov et al. (2002) proposed a bottom-up approach that resolves some of the gaps derived from the lack of understanding the speech and gesture production mechanism. Their approach relies on correlating speech prosodic features with deictic gestures by means of three different phases: *phonological*, *pragmatic* and *semantic*. At the semantic level, gestures and speech are synchronised to express the same idea; however, at the phonological level, different movements synchronise with different prosodic structures of speech, some of them discussed in Caldogneto et al. (2004).

We have so far discussed the gesture and speech annotation and coding, where gestures comprise mainly hand and arm movements. However, the multimodal interaction is usually carried out using other modalities such as gaze, facial expressions and 3D movements. Several approaches can be found in the literature to deal with the annotation of these modalities. For example, Schiel et al. (2002) proposed a Framework for Multi-Modal Resources (FMMR) to integrate signals via QuickTime, annotations via BPF (BAS Partitur Format), and linking them using physical time. According to the authors, the framework is appropriate to integrate these modalities since, on the one hand, QuickTime can integrate video, audio and images without changing their original format. On the other hand, the ASCII basis of BPF makes the annotation transportable and extensible to new tiers and platforms. According to the authors, the main drawback of this approach is the synchronisation of the different time bases in the original signals.

The annotations can refer not only to the semantics of the multimodal interaction by means of the interpretation of gestures, facial expressions or user utterances. In addition, they can be used to represent the user state in the course of the dialogue. Following this idea, Steininger et al. (2002) proposed an approach that encodes subjective user states taking into account the impression that a human has of the user state while interacting with the SmartKom system in a WOz session. During the annotation, the user states are encoded considering several categories to increase consistency and conformity between labellers. This classification is complemented by a prosodic labelling of the user utterances, thus speech is used to contribute together with the visual modality to represent the user state.

[7] http://www.mpi.nl/IMDI/
[8] http://www.mpi.nl/world/tg/lapp/eudico/eudico.html

5.2 Dialogue Modelling

Several theories have been proposed to model the dialogue between two interlocutors. One of the first was presented by Grice (1969). According to this theory, the dialogue partners must follow some principles to avoid dialogue breakdown in terms of quantity (each interlocutor must provide the information requested by the dialogue partner), quality (both partners must be sincere and not provide false information), relevance (they must provide information relevant in the dialogue), and mode (they must be clear, brief and concise).

Some years later, Clark and Schaefer (1987) presented the so-called *contribution model*. According to this model, dialogue is a collective activity in which communication problems can arise, as a result of the difference of knowledge of the interlocutors, use of different words, etc. Throughout the dialogue, the interlocutors interchange information and are able to detect and correct what can arise during the conversation. The Clark and Brennan (1991) called this process *grounding*. According to this idea, a dialogue is made up of *contributions*, which have two phases: *presentation* and *acceptance*. In the former, one of the participants in the dialogue utters a sentence for the interlocutor, whereas in the second phase the speaker waits until the interlocutor generates some evidence of comprehension of the sentence, which shows that the sentence is part of the common knowledge (*ground*) of both interlocutors in the dialogue. The acceptation phase can be as simple as a sentence, or something more elaborate if it contains a subdialogue for clarification or correction. In other words, each utterance U_n is connected with the precedent sentence U_{n-1}. With U_n the interlocutor provides evidence of comprehension of U_{n-1}. The receiver of U_n interprets the evidence: if considered as enough, proceeds with the dialogue, pronouncing U_{n+1} as the new relevant utterance. If otherwise, the evidence is considered either insufficient or negative, he uses U_{n+1} to initiate a correction subdialogue. Each interlocutor uses his own criteria to evaluate the comprehension evidence and determine whether or not to look for more evidence (which the authors called *grounding criterion*). This procedure is directly related with the confirmation strategies currently employed by dialogue systems (discussed in Section 5.3.2). According to this procedure, for example, a system that provides airplane travel information must provide the user with some evidence of understanding him. This evidence will be just oral in the case of SDS, or comprised of several output modalities in the case of MMDSs (e.g. speech and graphics). Both theories were developed to model the dialogue between human beings, but can also be applied to model the dialogue with dialogue systems.

The dialogue model obtained from the dialogue annotation must be implemented into the computer so that the dialogue system can interact with the user successfully to provide a specific service. Several approaches can be found in the literature to achieve this, which in general terms may be grouped into two categories: *finite-state* and *locally handled* models. In the former, the dialogue structure is represented as a state-transition network in which the states represent system prompts and the transitions represent the user utterances. Using this model, the possible next states in the dialogue for a given state are predetermined in advance following the transitions. In the locally handled models, the paths in the dialogue are not predetermined in advance and the interaction evolves dynamically, considering some kind of computation of the next *dialogue act*. There are several types of locally handled models such as plans, frame-based, object oriented, or event driven. Most commercial dialogue systems use state-transition models (or some variation of it); however, locally handled models have also been applied to commercial systems. For example, the SpeechMania toolkit (formerly by Philips, now by ScanSoft) uses event-driven models, whereas the DialogueBuilder (by Nuance) relies on the object-oriented models.

5.2.1 State-Transition Networks

State-transition networks allow modelling the user-system interaction in a very structured way, considering the data that must be exchanged at every stage of the dialogue (Aust and Oerder 1995). The advantage lies in their simplicity: the dialogue system moves along the network and obtains the necessary data from the user in order to carry out a particular task. Figure 5.2 shows a simple example for providing air travel information.

State-transition models have been criticised for their inadequacy in handling complex dialogues and their lack of flexibility. The possible *paths* in the dialogue are predefined

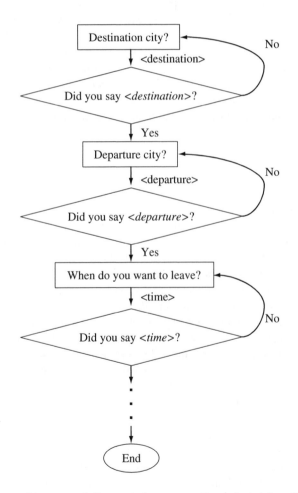

Figure 5.2   State-transition network for an airplane reservation (adapted from Heeman et al. 1998)

in the network, which makes it impossible for the user to deviate from them. There are problems when a user interaction is over-informative, i.e. he provides more information than requested by the system. For example, in the example shown in Figure 5.2 the system prompts the user for the necessary data items in the following order: *destination > source > time*. However, when the system prompts for the destination, the user may say 'to Boston ... leaving in the morning' or any combination of the other data items, which probably would not be correctly recognised since only the city name was expected. There are two possible solutions to this problem. One is restricting clearly the user responses so that he provides only one required data item. In order to provide more flexibility, a second solution is allowing a wider range of possible user answers in each state, and include additional transitions to transit to other states according to the data provided by the user. However, as long as the number of possible prompts increases, the number of possible transitions also increases, and allowing for the added flexibility increases it even more, which may create a problem in the setting up and maintenance. For example, it was estimated that adapting the Philips SDS (designed to provide train information) to handle flexible user responses, would require approximately that each system prompt could be followed by any other system prompt, thus leading to several tens of thousand of transitions (Aust et al. 1995). In general, state-transition networks are appropriate to model simple tasks that are clearly structured and require a few system prompt types, such as travel or bank account information. However, when the complexity increases, these models tend to be less appropriate. For example, this is the case of the Trains system (Allen et al. 1996) with which the user interacts to build a plan incrementally, incorporating new restrictions as the dialogue proceeds.

5.2.2 *Plans*

Dialogue modelling based on plans aims at resolving the problems derived from the lack of flexibility of state-transition networks (Allen et al. 1995). This model relies on the observation that human beings plan their actions to achieve determined goals. A dialogue system that employs this model must be able to infer the user goals as well as build and activate plans to provide the service requested by the user. Plan-based systems have been studied in the field of Artificial Intelligence. Their mathematical foundation is inference, which is applied to a set of logical rules and axioms that represent the knowledge about the application domain. Using plans, the system interacts with the user to achieve some facts, the achievement of these facts triggers rules, and these rules can create new facts that again must be achieved by the system. A plan-based model can be observed as a state-transition network in which the states are generated dynamically and are not limited to a predetermined finite set. The cability of the system to process an unlimited number of states is one of the advantages of this approach in terms of scalability.

The creation of plans requires carrying out quite deep studies of the rational behaviour of the dialogue partners, which allows for a more flexible interaction than provided by state-transition networks. However, building and using plans can be very complex in practice due to the dependence on the inference process, which is very difficult to model. In general, complex application domains require a great deal of effort to create the inference rules.

5.3 Dialogue Management

The success of a dialogue system depends fundamentally on the correct design of the strategies to interact with the user, in such a way that they avoid, as long as possible, the current limitations of the technologies employed (e.g. low ASR recognition rates in noisy conditions). The interaction strategies must be designed in such a way that the user can obtain the requested service in an acceptable time, and through an interaction as comfortable as possible. If both requirements are not met, the system will be considered useless. The dialogue manager (Figure 1.1) is the system module that drives such an interaction. Among other tasks, this module must interaction take the adequate actions so that errors that may occur during the dialogue can be corrected.

5.3.1 Interaction Strategies

In the context of dialogue systems, the dialogue is a collective activity carried out in real time by a human being and a computer program that has limited computational and knowledge resources. Throughout the dialogue, both agents must collaborate to make the interaction successful. This collaboration not only is necessary in the case of human-computer dialogues but also when both dialogue partners are human beings. Even when both humans share the same physical location, speak the same language and use common words, there is no absolute guarantee that one will understand the other. In fact, one may overestimate the other's knowledge, may not hear what the other says, etc., thus making the collaboration between both necessary for mutual understanding. As commented above, Clark and Brennan (1991) called this collaborative process *grounding*, according to which, each dialogue interaction contributes to a higher-order structure that both participants must try to achieve. Considering the initiative control made by the participants during the dialogue, three interaction types can be considered (Levin et al. 2000):

- *User-directed interaction.* When this interaction type is used, the user always has control over the conversation throughout the dialogue, and the system just responds his queries and orders. This interaction type is not commonly used nowadays since it typically leads to a poor system performance. This happens because the user does not rely on the system guide, and thus believes himself to be free to say whatever he wants, which tends to cause ASR and speech understanding errors.
- *System-directed interaction.* When this interaction type is used, the system always has control over the conversation throughout the dialogue, and the user just answers its prompts. The advantage of this strategy is that it tends to reduce the possible user inputs, which typically leads to very efficient dialogues since all the system goals are generally achieved. The disadvantage is the lack of flexibility on the part of the user, who is restricted to behave as the system expects, providing the data, if necessary, in the order specified by the system. Typically, these systems allow the user to use some commands to navigate explicitly in the dialogue flow and carry out actions such as going back one step, start over, ask for help, etc.
- *Mixed-initiative interaction.* When this interaction strategy is used, both the user and the system can take the initiative in the dialogue, consequently influencing the conversation flow. The advantage is that it combines the advantages of the two previous strategies. On the one

hand, the system guides the user and thus reduces the possible out-of-task utterances, while, on the other, the user can take the initiative and provide data items not explicitly requested by the system (i.e. be over-informative). The user can also take the initiative and interrupt the system to correct data provided in previous turns. This interaction type requires facing the problem of dealing with greater vocabularies, and also with greater diversity of possible sentence types and user intentions at every dialogue state.

From these descriptions, one may think that the mixed-initiative interaction strategy is always the best, regardless of the task to be performed by the system. However, several studies indicate that this is not necessarily true. For example, Hone and Baber (1995) studied the relation between dialogue restrictions and time necessary to carry out the tasks. They concluded that dialogues carried out using the system-directed strategy, with yes/no confirmations for all the data provided by the user, require great task-completion times. However, the authors also indicated that this time depended mainly on the performance of the speech recogniser; they observed that when they used a less restricted interaction strategy, the error probability was greater, which increased the time necessary to complete the tasks due to the existence of a greater percentage of correction turns. In the same direction, Potjer et al. (1996) studied the performance of two versions of the same dialogue system: one used system-directed interaction strategy, isolated speech recognition and NLP based on keyword detection, while the other used mixed initiative, continuous speech recognition and a complex NLP. Using the first version, the system prompted the user for the necessary data to perform the task in two steps, whereas in the second the data were requested in just one step. In principle, it seems reasonable to think that the required number of turns would be smaller it the mixed initiative was used. However, in practice this strategy required additional turns to correct ASR errors. According to user tests, the first version of the system was not considered to be slower than the second; a subjective test showed that they were satisfied with both versions of the system.

Taking into account the diverse studies carried out, it could be said that the selection of the most suitable interaction strategy for a dialogue system depends mainly on the task to be performed by the system. In general, typical tasks for which it is preferable to use the mixed-initiative interaction strategy are those in which the data items to provide the requested service may be provided by the user in any order (e.g. travel information and reservation). On the other hand, for applications such as directory information or stock exchange information, it can be very easy to determine that the user has uttered a sentence of type 'I want to know the telephone number of <name>'; the problem is to recognise correctly the *name*, which may be any in a list containing several hundred thousand entries. For this type of application that still represents a challenge for ASR technology due to the use of very large vocabularies, with very similar likely words, it may be preferable to use a more restricted interaction strategy (i.e. system-directed) in order to enhance ASR recognition rates and thus the task completion rates. Some systems use system-directed or mixed interaction following an adaptive strategy; they start using mixed-initiative by default and if the dialogue becomes unsuccessful then change automatically to system-directed.

5.3.2 Confirmation Strategies

Considering the current limitations of the technologies employed to build a dialogue system, it is necessary to suppose that the information obtained from the user input may be incomplete, inaccurate or inconsistent with previous information (e.g. because of ASR errors, words can be inserted, deleted or substituted). To avoid these errors causing problems in posterior analysis stages, dialogue systems use confirmation strategies to confirm determined words, or complete sentences. Usually, two types of confirmation strategy are used, called *explicit* and *implicit*. When the former is used, the system generates an additional dialogue turn to confirm the data obtained from the user in the previous turn. As can be observed in the following example, the disadvantage of this strategy is that dialogue tend to be lengthy due to the additional confirmation turns (U2 and U4), which makes the communication less effective and even excessively repetitive if all data provided by the user are confirmed (S = system, U = user).

S1: Destination city?
U1: Madrid
S2: Did you say Madrid?
U2: Yes
S3: Departure city?
U3: Tokyo
S4: Did you say Tokyo?
U4: Yes
S5: Day of the week you want to leave?
U5: Monday
. . .

On the contrary, if the implicit confirmation strategy is used, no additional turns are necessary since the system includes the data to be confirmed in its next dialogue turn. In this case, it is responsibility of the user to make a correction if he observes a data item wrongly obtained by the user. This confirmation strategy can be observed in the following example dialogue, in which the user confirms implicitly (by turns U2 and U3) the data obtained by the system in turns U1 and U2, respectively.

S1: Destination city?
U1: Madrid
S2: Ok, destination Madrid. Departure city?
U2: Tokyo
S3: Ok, departure from Tokyo. Day of the week you want to leave?
U3: Monday
. . .

Explicit confirmations tend to increase the number of necessary turns but facilitate the error correction on the part of the user. On the contrary, implicit confirmations tend to reduce the necessary turns, but it may be more difficult for users to correct the errors.

5.3.2.1 Use of Confidence Scores

The literature shows a diversity of confidence measures for the recognised sentences and words, which can be obtained from acoustic and language models jointly, or from any of them separately (Rüber 1997; Williams and Renals 1997). The use of these measures, together with one or more confidence scores allows the dialogue manager to decide whether to discard the recogniser output, accept it or confirm some words. One possible way to compute the confidence scores is based on the use of the N-best list generated by the speech recogniser, and suppose that the words appearing in many hypotheses have a higher probability of being correct, and thus must have high confidence scores. Souvignier et al. (2000) and Schramm et al. (2000) used a confidence score based on this approach; concretely, the confidence score of a word A is computed by summing the probabilities of the hypotheses in which this word appears. To make the computation, I_A defines the set of indexes of hypothesis containing A, while sc_i represents the probability of the hypothesis i in the N-best list. Taking into account this notation, the confidence score of the word, $C(A)$, is computed as shown in the following formula,

$$C(A) = \frac{\sum\limits_{i \in I_A} e^{-\lambda \cdot sc_i}}{\sum\limits_{i=1}^{N} e^{-\lambda \cdot sc_i}} \tag{5.1}$$

where the λ factor is used since the scores provided by the recogniser are scaled negative logarithms of probabilities. This factor is used as a tuneable parameter that decides how the probability mass is distributed over the N-best list. Several confirmation strategies can be defined based on the confidence scores. For example, a possible strategy confirms only the words whose confidence scores are below a determined confidence threshold. A second strategy explicitly confirms a word if its corresponding confidence score is not higher than a threshold, and confirms it implicitly otherwise. A third method, used in many dialogue systems, relies on using two confidence thresholds: C_1 and C_2; C_1 represents a high confidence[9] threshold (e.g. 0.8) and C_2 represents an intermediate threshold (e.g. 0.5). When deciding whether to confirm a word, the word confidence score is compared against the thresholds and a decision is made on this comparison. If the word confidence score is greater than C_1, then the word is considered to be correctly recognised and is not confirmed. If the score is smaller than C_1 and greater than C_2 then the word is implicitly confirmed. Finally, if the score is smaller than C_2, then the word is explicitly confirmed. This confirmation strategy achieves more efficient dialogues since it avoids confirmation turns for the words that, in principle, are considered to be correctly recognised.

The main problem of using confidence scores to decide the confirmation strategy is the so-called *false alarms*, which can be of two types: false acceptation and false rejection. The former occurs when the confidence score of a wrongly recognised word is greater than a threshold, where as the second occurs when the confidence score of a correctly recognised word is below a threshold. A false rejection is not excessively prejudicial since it only conveys an additional (unnecessary) dialogue turn to obtain again the word from the user. However, a false acceptation can be disastrous for the system since it means a wrongly recognised word is considered correct. If confirmation turns are saved for words with high confidence score, and the word's score is greater than the threshold, the word will not be confirmed, and thus it will be impossible to correct it (if no additional confirmations are considered).

---

[9] In this example, it is assumed that confidence scores are real values in the interval (0, 1).

5.3.2.2 Adaptive Strategies

The performance of a dialogue system can vary notably from one user to another, and even can vary for the same user in the current interaction. In order to take this fact into account, several methods have been proposed to adapt the interaction and confirmation strategies to the problematic situations that may occur in dialogues when it becomes difficult for the system to understand the user. For example, Litman and Pan (2000) presented an adaptive technique implemented as a set of *if-then-else* rules that determine problematic situations in the dialogue. Using this technique, if a problematic situation is detected, the system adapts the confirmation and interaction strategies, which implies stop using default non-restrictive and user-comfortable strategies, e.g. implicit confirmation and mixed-initiative interaction, and then start using more restrictive strategies, e.g. explicit confirmation and system-directed interaction. The rules are trained using a corpus of dialogue labelled as 'good' or 'bad' considering five parameters: acoustic confidence, dialogue efficiency (measured as the number of system turns), quality and naturalness of dialogues (measured as the number of user turns employed to ask for help), experimental parameters (e.g. initial configuration of the interaction strategies) and lexical parameters in the system output (i.e. lexical items in the output of the speech synthesiser). The trained rules are obtained by analysing the tagged dialogue corpus with the RIPPER program (Cohen 1996).

5.4 Implications of Multimodality in the Dialogue Management

5.4.1 Interaction Complexity

The dialogue management carried out by MMDSs is much more complex than that carried out in SDSs due to several reasons. The first is that multimodal systems must take into consideration not only one but several input modalities that can be used to interact in different ways with the system. For example, in a system developed for the ATIS domain that allows interaction using speech and pointing on a PDA screen, a user can say 'Flights from', select 'Pittsburgh' by pointing, then say 'to' and select 'Boston' by pointing; or he can say 'Show me the flights from Pittsburgh to this city' while he points to 'Boston' to specify the meaning of 'this city'. Given that the diverse input channels are not reliable, the dialogue manager must employ more robust dialogue strategies to face the errors than in the case of SDSs. One advantage of using several input modalities is that in case of errors, the dialogue manager can suggest the user changes the interaction modalities used to input data (e.g. use pointing only) taking into account the environmental conditions (e.g. silent or noisy, luminous or dark, etc.). This requires the system to keep track of the modalities used to provide each information chunk; for example, Denecke and Yang (2000) used multidimensional TFSs to store not only the data provided by the user but also the modalities used for doing so. Using one device or another to provide the input may be transparent for the dialogue management if every recogniser uses the same format to represent the recognised data (e.g. an XML-based language as studied in Section 2.2.2).

A second reason is that SDSs are developed to be used in contexts in which the conversations are very predictable and restricted, as is the case, e.g. in the ATIS domain. Although other more elaborated dialogue models can be set up, for this application domain it suffices to use a state-transition network in which each state represents a prompt the system must generate to obtain specific data from the user (e.g. the destination city) and each transition represents a dialogue move. The complexity of such a dialogue is very low in comparison with the one required for new applications supported by multimodal systems. An example is the WITAS system (Lemon et al. 2001) developed to dialogue with mobile autonomous robots that have their own perception of a real and changing world with which they interact. The dialogues that can be carried out with this type of system are clearly different to those that can be carried with a travel information system. For example, there is no clearly pre-defined sequence of events in the dialogues, the robot may 'need' to communicate with its operator at any time, there may not be clearly defined states concerned with the termination of dialogues, the relevant discourse objects in the environment may appear and disappear dynamically, etc. Clearly, the dialogue modelling and strategies developed for the ATIS domain are not adequate to deal with this type of application.

A third reason for more complexity in the multimodal dialogue management is the need to handle not only the semantic representations obtained from the user inputs but also the semantic representations obtained from the system outputs, which must be compatible at the dialogue management level. In other words, in addition to understanding the user input, multimodal systems must also understand their own outputs. This is necessary to resolve cross-modal references since the user may make an oral reference to a graphic object the system previously placed on the screen, for example. To resolve the reference, the object must be internally represented in a language compatible with that used to represent the user spoken input so that both can be integrated into the fusion process. For example, the SmartKom system (Wahlster 2002) uses a module that continuously creates M3L representations for the objects placed on the screen, which allows them to be included in the dialogue context and resolve anaphoric, cross-modal and gestural references.

A fourth reason is that allowing several interaction modalities implies the dialogue system must be robust enough to handle a wider range of potential users than SDSs, who may differ largely in several aspects, for example, experience using these systems, age, preferences, ability, sensory or motor impairment, etc. User modelling can be very useful to differentiate user types and adapt the system accordingly. This is especially useful in the case of mobile devices (e.g. PDAs), which are generally of personal use. User modelling information (e.g. needs, preferences and/or abilities) can be stored in user profiles, which can be used to adapt the system and enhance its performance in advance for each user. For example, in the case of speech impairment users, the user profile may indicate the system must use gestures to augment or substitute speech input. Taking into account this modelling, the dialogue management must be adaptable to provide more feedback to inexperienced users, whereas users who make repeatedly specific types of errors can receive specific advice on using the system. The dialogue manager must also support interaction strategies adapted to the different user types (e.g. system-directed initiative can be used for inexperienced users whereas mixed-initiative can be used for experienced users). Moreover, in situations in which the user wants to keep his privacy, the dialogue manager must use the convenient modalities to avoid third people listening into the conversation.

A fifth reason is in terms of the interruptions made by the user during the system output, which generally occur due to changes of requests, corrections or changes of mind. Steininger et al. (2001) classified the interruptions into several categories: *abort* to stop the current processing/presentation (e.g. 'Stop'); *premature request* to abort the presentation and fulfil the request (e.g. 'ok, this one'); *correction* to change the request being processed (e.g. 'no, the right one'); *successive request* to include new information in the request being processed (e.g. 'and another one too'); and *back-channelling* to provide feedback about the correct understanding (e.g. 'ok'). In SDSs the interruptions can only be made using speech (and possibly DTMF), but in a multimodal interaction the user can interrupt the system using a variety of modalities. For example, the output of the Rea system (Cassell et al. 1999) is interrupted if the user starts to speak (typical barge-in in SDSs) and also if he makes gestures associated with the turn taking behaviour.

To deal with the specific problems of MMDSs and make the interaction with the users as natural and effective as possible, specific dialogue models have been developed. For example, Lemon et al. (2001) used a set of rules for the WITAS system to interpret the multimodal input generated by the operator and the robot. To allow for a robust dialogue management, the dialogue model uses a tree structure of dialogue states in which edges represent dialogue moves and branches represent dialogue threads. The system also uses a hierarchical task-tree that represents the operator tasks and sub-tasks, as well as task reordering and reference (e.g. the operator can say 'Go to the tower. Show me car 1. Actually, do that first'). In the MATCH system (Johnston et al. 2002), the dialogue management is based on a set of rule-based processes that deal with a *shared state*, which represents the intentions and beliefs of the system and the user, the dialogue history, a focus space, and information about the user, the domain and the available interaction modalities. Using these information sources the system carries out the interpretation of the user input, obtaining the most likely hypothesis, taking into account the shared state, the determination of the next possible system action, and the selection of one of them, which updates the system's model of the user's intentions as a result. Beskow and McGlashan (1997) used an event-based strategy for the dialogue management in the Olga system. Depending on the events generated by the user input, the system updates an interaction model consisting of the system goals and a set of interaction strategies. The goals determine the behaviour of the system allowing for confirmation and clarification of the user inputs to minimise the dialogue breakdown, as well as prompting for additional information to maximise the dialogue progress.

5.4.2 Confirmations

Given the current limitations of the recognition technologies, dialogue systems must confirm the data obtained from the user. Typically the confirmation is based on confidence scores (associated with the recognised words, in the case of SDSs). In multimodal systems the confirmation can be made either at an early stage (for each modality before the fusion process) or at a late stage (after the fusion process) (McGee et al. 1998). In SDSs the dialogue manager uses the confidence scores and decides whether to confirm words as well as the confirmation method (typically either implicitly or explicitly). Given that multimodal systems can use several input modalities, several confidence scores can be considered, which can make the confirmation strategy more robust and efficient. For example, in a system designed for the ATIS domain, the user may provide a destination city using speech, pointing or both modalities at the same time, i.e. he can say 'I want to travel to Boston' while he points to the city on a PDA screen, thus obtaining two confidence scores, one for each modality. If both scores are higher than a threshold and the semantic interpretations from both modalities match, it can indicate the destination city has been correctly obtained, and thus the confirmation can be avoided or carried out implicitly.

Additionally, in SDSs all the confirmations must be carried out using speech, which can make the interaction tedious and unnatural. Multimodality can make the confirmation strategy more natural and informative, since the system has many ways to provide feedback to the user. For example, the MATIS system (Nigay et al. 1993) uses a multimodal confirmation strategy based on confidence scores that works as follows: words with confidence scores below a threshold are shown on the screen and confirmed explicitly, whereas words with confidence scores higher than the threshold are shown on the screen but are not confirmed. This confirmation strategy, based on two modalities (speech and graphics) is faster and more informative for the user. The AdApt system (Gustafson et al. 2002), developed to provide apartment information, shows on the screen the restrictions currently used to query the database, which provides a graphic confirmation of the data obtained from the user (e.g. the apartment price must be lower than a spoken amount of money). In the case of the Olga system (Beskow and McGlashan 1997), developed to provide information about microwave ovens, the confirmation strategy is performed using speech, graphics and gestures made by the system's animated agent. The success in understanding the user input is indicated by a nod gesture, whereas a misunderstanding is transmitted through a spoken explanation accompanied by the agent's raised eyebrows and the mouth turned down. If the database query does not provide information for the user, a spoken informative message is generated, accompanied by a regret facial expression by the agent. The information obtained from the database is communicated using speech and graphics, whereas detailed information about a particular product is transmitted graphically while the agent generates a spoken summary.

5.4.3 Social and Emotional Dialogue

As discussed in Section 3.2.5, a line of research within the field of MMDSs is concerned with the setting up of empathic system-user relationships, as an attempt to reduce the negative feelings the user may experience during the interaction, such as frustration or face threat. To achieve this goal, the systems must employ more sophisticated dialogue management strategies than the ones used in task-only oriented systems. Several approaches to the problem can be found in the literature. For example, in the Rea system developed for the real estate domain (Bickmore and Cassell 2004), the dialogue manager interleaves small task and task talk during the initial interview with the user. The system turns are decided in such a way that minimises the face threat to the user and maximises trust, so that the system can achieve its goals in the most efficient manner possible. The dialogue management is carried out using an activation-based approach that allows smooth transit from deliberative, planned behaviour to opportunistic, reactive behaviour, allowing as well several, non-discrete goals to be pursued. Each node in the network represents a system decision in the dialogue. The system decides to do small talk when it needs to increase proximity to the user, for example, before prompting a task question that may be unpleasant for the user to answer (e.g. the maximum price he is willing to pay for the apartment). Small talk is also used to move the dialogue focus little by little to the particular issue the system is willing to address. One advantage of using the activation network is that a simple adjustment of some parameters makes the system more or less coherent, more or less polite, more or less task-oriented, etc.

Another line of research is concerned with the recognition of the emotional state of the user and generation of emotional responses. Emotion is used as an additional modality in order to make the system behaviour more human-like. In SDSs, emotional information can be obtained considering the acoustic realisation (e.g. talking speed, volume, pitch, energy, etc.) as well as the syntactic and semantic form (e.g. use of vocabulary or affective phrases). Multimodal systems can obtain emotional information using additional input channels (e.g. user bio-signals, facial expressions and gestures). Thus, the processing of emotions implies the dialogue manager can correctly integrate and interpret emotional information together with the semantic information obtained from the user. Several methods can be found in the literature to combine both types of information. For example, Holzapfel et al. (2002) used a dialogue manager structured into three levels. The lower one works at the sentence/grammar level to decide the next goals and prompts of the system. The middle level handles an abstract representation of the dialogue status, and decides the dialogue strategies. Finally, the higher level contains meta-strategies that take into account the emotional information obtained from the user to decide the middle-level strategies that must be used during the dialogue. The emotional information is stored in multidimensional TFSs, and is used to determine whether the user is in a *neutral* emotional state, or he was *happy*, *stressed*, or *angry*. Another example is the 'affective reasoning engine' used in the PPP Persona system (André et al. 1999) to handle two channels of emotional information: primary emotions (e.g. being frozen by terror) generated using simple reactive heuristics, and secondary emotions generated by the affective reasoning engine using the OCC emotional model (Section 3.2.5.1). Emotions are mainly used to determine the system's short-term emotional state, which is expressed using gestures and speech inflection, but they also influence the selection of the sentences the system generates during the dialogue, and are also used to adjust the personality features of the system.

5.4.4 Contextual Information

As discussed in Section 3.1.1, one of the advantages of MMDSs is that users can express themselves in a more natural way. However, this implies many potential problems for the systems given the great range of possible inputs, which will not always be made using the best modalities for the current environmental conditions (e.g. users may use speech instead of pointing in noisy environments) causing uncertainty, i.e. lack of confidence in the interpretation of the input. Moreover, the diverse analysis modules (e.g. speech recogniser, gesture recogniser, etc.) can produce different interpretations for a given input, thus causing ambiguity (equally likely hypotheses of the input). The resolution of the ambiguity in selecting the appropriate hypothesis is a crucial issue when guessing the intention of the user. To do so, the dialogue manager can enrich the diverse hypotheses with contextual information in order to select the correct one. To do so, several researchers have used the *unification* operation (Lascarides et al. 1996) which is a binary and commutative operation typically applied to combine information chunks provided by several modalities that do not contain conflicting parts. When the information chunks contain conflicting parts, which can be originated when the user changes his mind (e.g. the data he previously gave to the system to attend a movie show), Alexandersson and Becker (2001) proposed the *overlay* operation to enrich the information obtained from the user with some kind of information inherited from the dialogue context. Overlay is a binary and non-commutative operation that rewrites the conflictive information. Informally, it can be assimilated to putting shapes on each other, as can be observed in Figure 5.3, where the covering refers to the new information provided by the user and the background to the old (previous) one. The result of this operation is the rewriting of the conflictive information in the background with the information in the

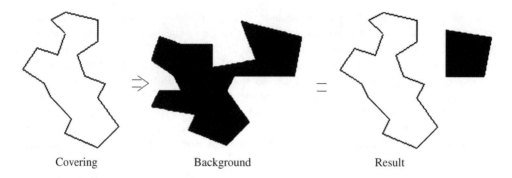

Covering   Background   Result

Figure 5.3 Overlay is 'Putting shapes on each other' (reproduced by permission of Alexandersson et al. 2001)

covering. In essence, this operation uses the unification operation if there are no conflicts in the information to combine, and otherwise uses the information from the first argument.

The authors also proposed the following scoring function for the overlay operation using TFSs (Section 2.2.2.4), which denotes how well the covering fits the background regarding non-conflicting features:

$$score(co, bg, tc, cv) = \frac{co + bg - (tc + cv)}{co + bg + tc + cv}$$

This function uses four parameters, initialised to zero, that are increased during the computation of the overlay: co (a feature or atomic value that occurs in the result of the overlay stems from the covering), bg (a feature or atomic value in the result of the overlay stems from the background), tc (type clash, i.e. the types of covering and background are different) and cv (value conflict, which happens when the value of a feature in the background is overwritten). The value of the positive extreme of the interval ($score(co, bg, tc, cv) = 1$) indicates the TFSs can be unified, whereas the value of the negative extreme ($score(co, bg, tc, cv) = -1$) indicates all the information in the background has been overwritten by the information in the cover. Any other value in the interval indicates the cover more or less fits the background (the greater the value, the better the cover fits the background).

Both operations (unification and overlay) have been used in the SmartKom system (Wahlster et al. 2001; Wahlster 2002), which features an architecture for dialogue management that resembles the standard architecture of SDSs: the dialogue manager of Figure 1.1 is divided into two modules: *action planner* and *discourse modeller* (Pfleger et al. 2003). Each recognition module in the system produces a hypothesis and a score associated with it, and so does the discourse modeller. For example, the score produced by the natural language processing module (Figure 1.1) indicates how well the sequence of words it receives can be parsed and transformed into a semantic representation. Considering the scores produced by the different recognisers and analysis modules, the discourse modeller decides the most likely hypothesis, which is sent to the action planner module. The discourse modeller also produces a score for that hypothesis that indicates how well it fits the context. For explanation purposes, let us consider this dialogue excerpt with the SmartKom system (reproduced by permission of Pfleger et al. 2003), in which [↑] denotes a deictic gesture made by Smartakus, the animated agent of the system:

U1: I'd like to see a movie tonight.
S1: [Displays a list of movies] Here [↑] you see a list of the films running in Heidelberg.
U2: Hmm, none of these films seems to be interesting . . . Please show me the TV program.
S2: [Displays a list of films] Here [↑] you see a list of broadcasts running tonight.
U3: Then tape the first one for me!

During the analysis of U2, the dialogue manager receives a set of hypotheses that are compared and enriched with information from turns U1 and S1. Although U2 and U1 have different topics (U1 is concerned with cinema movie schedules whereas U2 is concerned with TV movie schedules) the temporal restriction (*tonight*) from U1 is transferred to the interpretation of U2 since it is in the context.

5.4.5 User References

In a multimodal interaction the user has more ways to make references, not only to others words but also to other entities that take part in the dialogue, such as icons or pictures on the screen. Thus, on the one hand, the dialogue manager must face the problem of resolving the classical reference types (*endophora* and *deixis*) (Brøndsted 1999). The endophora is the reference using a linguistic entity to another linguistic entity, usually mentioned before (anaphora) but that can also be mentioned later (cataphora). The deixis is the reference using a linguistic entity whose interpretation is relative to the extralinguistic context, e.g. who is speaking, the place or moment of speaking, etc. On the other hand, the dialogue manager must face two additional reference types: *cross-media* and *cross-user/system* references. The first type is the (deictic) reference using a linguistic entity to an antecedent in other communication channel, e.g. a picture or a video shown on the computer screen. The second type is the endophoric/deictic reference of a linguistic entity in the user input to an antecedent in the system output or vice versa (a reference in the system output to an antecedent in the user input).

The resolution of these reference types requires the system to understand its own output the same way it understands the user input. This is in clear contrast to SDSs, which are generally focused on understanding the user input and just generate responses that the system does not need to understand at all. For example, let us suppose a multimodal system to provide information about city facilities, e.g. hotels, restaurants, etc. shown on a map on the computer screen. If at a moment of the dialogue the user says 'Show me the photo of *this* hotel', the dialogue manager must resolve the reference made by the linguistic unit *this*. To do so it can start by waiting for a deictic gesture for a determined time (e.g. 2 or 3 seconds). This time is necessary to accommodate the different synchronisation of speech and gestures, given that different users (or even the same user in different situations) can make the gesture before, during or after speaking. Also, the response time of both recognisers is typically different. If no gesture information arrives, the dialogue manager can consider the dialogue history and contextual information, and use a 'salience list' to find a possible referent, taking into account several factors, as for example, gender and number correspondences, semantic consistency, semantic and syntactic parallelism, proximity, etc. If no hotel name is found in the list, the dialogue manager should then ask the user. If several hotel names are shown on the screen and the user answers 'The one on the up right corner', the dialogue manager must use the knowledge about the screen content, which should have been 'understood' by the system when it placed the different hotels on it. Using this information, represented adequately to be combined with the user spoken input, the dialogue manager can resolve the reference. This procedure has been used by several systems, for example, WITAS (Lemon et al. 2001), which uses a 'salience list' associated with the dialogue state that consists of objects referenced in the dialogue thus far, ordered by recency. This list includes information about *how* the references are made (i.e. by which modality), which is important for resolving and generating anaphoric and deictic expressions in the dialogue.

Other authors have proposed layer-based architectures to deal with contextual information and multimodal references. For example, the Smart Sight system (Yang et al. 1999) uses a four-layer architecture that contains the input representations (such as text or gesture hypotheses), parse trees, semantic representations, and objects that referring expressions might refer to, respectively. The SmartKom system (Pfleger et al. 2003) uses a three-layer architecture for modality, discourse and domain objects, respectively, as shown in Figure 5.4.

In this architecture, an object in the Modality Layer contains information regarding the modality used either by the user or the system to introduce this object into the context. Three types of object can be found in this layer: linguistic objects (LOs) introduced by spoken expressions; visual objects (VOs) introduced by graphic presentations; and gesture objects (GOs) introduced by gestures. Each object in this layer is linked to an object of the higher layer: the Discourse Layer. A discourse object (DO) represents a concept that can be referenced by a referring expression. This concept can be either objects, events, states or collections of objects. Finally, the Domain Layer contains associations between objects in the Modality Layer and instances of the domain model. The local focus stack contains pointers to the DOs that can be possible referents of referring expressions (ordered by salience). A DO representing a collection of objects contains a partition that is created considering perceptive information (e.g. a list of movies shown on the screen) or discourse information (e.g. 'Do you have more information about the first and the second movie?', in the context

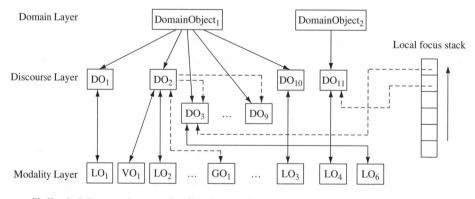

S2: Here [pointing gesture] you see a list of broadcasts running tonight.   U3: Then tape the first one for me.

Figure 5.4 Layer architecture to deal with contextual information and referring expressions in the SmartKom system (reproduced by permission of Pfleger et al. 2003)

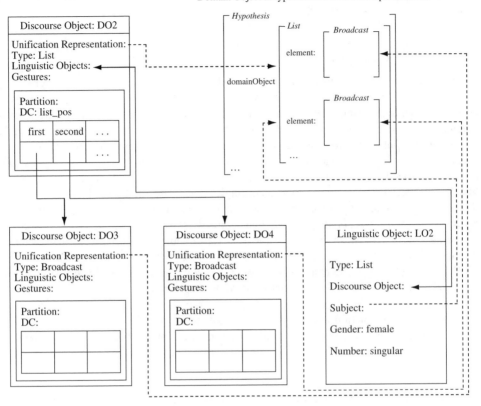

Figure 5.5   Discourse objects (reproduced by permission of Pfleger et al. 2003)

of the movies shown on the screen). Each element in a partition is a pointer to other DO, as can be observed in Figure 5.5, in which the discourse object DO2 represents the list of movies shown on the screen. This DO was put into the discourse layer by the gesture made by the animated agent in the turn S2 of the dialogue excerpt shown in Section 5.4.4: '*Here* [↑] *you see a list of broadcasts running tonight'*. Figure 5.5 shows the configuration of the DO2 discourse object with a partition.

The analysis of turn U3 in the previous dialogue excerpt requires the system to resolve the referring expression 'the first'. Given that there is no deictic gesture made by the user to be combined, the system must try to resolve the reference by accessing the contextual information. The resolution process is initiated by the data fusion module, which infers from the contextual information that the referent must be a domain object of type 'Broadcast'. Also, it infers from the definite expression ('the first') that the referent must be extracted from a partition. Thus, the fusion modules sends these information to the discourse modeller (a submodule of the dialogue manager) indicating as well that the partition differentiation criterion (DC in Figure 5.5) must be 'list_pos'. The discourse object DO2 is the first one that satisfies this requirement. Its partition element for the value differentiation criterion 'first' is a pointer to the discourse object DO3. Given that this object satisfies the requirement

'Broadcast' specified by the fusion module, it is the object considered to be the referent and is sent back to the fusion module.

5.4.6 Response Generation

As shown in Section 2.2.6, in the conceptual MMDS we are using in this book (Figure 1.1) response generation is carried out in the response generator module. This module receives abstract semantic representations from the dialogue manager (e.g. written in a XML-based language) and transforms them into the multimodal response of the system. This module decides the output to be produced through each output modality and coordinates the output across the modalities so that the output is synchronised for the user. It also takes into account the specific characteristics of the devices (e.g. screen size, wireless communication bandwidth, etc.) to adapt the output accordingly. Therefore, in principle we should not discuss the effect of multimodality on the response generation here since this section is concerned with the effect on the dialogue management. However, as discussed in Section 1.3, we could consider the response generator as a sub-module of the dialogue manager that decides the output generation. For example, in the case of the Olga system (Beskow and McGlashan 1997), it is the responsibility of the dialogue manager to do the work of the response generator. Therefore, it makes sense to discuss here the effects of multimodality.

Clearly, multimodality affects the response generation, which is much more complex in a multimodal interaction because several output modalities are available. Therefore, in a multimodal system there must be a module (dialogue manager or response generator) that decides the most adequate output modalities to generate a response (e.g. language or graphics) and the specific media to generate the response (e.g. language can be either written or spoken). This module must also take into account the features of the output devices (e.g., screen size, available bandwidth in mobile systems, etc.) as well as the environmental conditions (silent vs. noisy) to adapt the response accordingly, and must synchronise the output if the response is realised through several modalities. If the selected modality is only speech, the messages must be brief and concise in general to make the system reactive and natural, since the users do not like, in general, long responses, especially in the case of telephone interaction. However, if the system decides to use the display, then it can show long messages, graphs and tables with a large amount of data. Since all this information will be presented graphically to the user, it will remain onscreen and the user will be able to understand it better.

Multimodality provides more feedback to the user about the interaction status. For example, an animated agent can acknowledge user presence by turning its posture to face him, can nod its head or emit a short statement such as 'I see' in response to short pauses in the user speech, may raise its eyebrows to indicate partial understanding of sentences, etc. For example, the dialogue manager of the Olga system (Beskow and McGlashan 1997) uses a set of rules to decide the output modalities for the response generation, which take into account the system goals as well as the information type to provide to the user. To express a successful understanding, the system uses gestures; to express understanding errors, it uses speech and gestures; to express a relaxation of restrictions when accessing the database, it uses speech and gestures; and to provide information to the user, it uses speech, graphics and gestures. Moreover, multimodality means the user can recognise the system emphasis on particular words, which is very important to determine the key parts of the sentences. SDSs can emphasise particular words only by prosodic means (pitch accents), whereas multimodal systems can accompany words with beat gestures made with the head and/or the hands. The multimodal output can also help in the turn-taking function by providing visual information; for example, a microphone icon shown on the screen can be used to signal the user turn, whereas the animated agent can show a facial expression, look at the user or make a gesture to encourage him start/continue talking.

Additionally, if the dialogue system deals with emotion recognition and generation, the dialogue manager must also decide how to express the emotions in a manner easily interpretable to the user to obtain the desired effect. This is very important since, as pointed out by several researchers, personality and emotions can be conveyed in various ways. For example, Furnham (1990) indicated that extravert characters use more direct and powerful phrases than introvert characters, and Scherer (1979) showed that these characters speak louder and use more expansive gestures than introverts. Several methods can be found in literature to transmit the personality and system emotion to the user. For instance, in the case of the PPP Persona system (André et al. 1999), the dialogue manager annotates the actions to be executed with appropriate mark-ups in terms of personality traits (e.g. extraversion, agreeableness, neuroticism, etc.) and emotions (e.g. happy, sad, afraid, etc.) that are properly interpreted by the response generation module to generate and coordinate speech, facial expressions and body gestures accordingly. The system starts the interaction as an extravert, agreeable and emotionally-balanced character that takes the initiative, is cooperative and remains patient if the user asks the same question over and over again.

5.5 Implications of Multilinguality in the Dialogue Management

In this section we discuss dialogue management problems derived from multilinguality. At first glance, it would seem that there are no problems if the system handles the multilingual input using the interlingua and dialogue-control centred approaches explained in Chapter 4. Doing so, the input for the dialogue manager is interlingua (the language dependency is assimilated by the preceding modules), and there remains no linguistic content to drive the controller module. However, there still remain some language-dependent dialogue management issues, such as the reference resolution, ambiguity of the speech acts and differences in the interactive behaviour of the MLDSs, which are discussed in this section.

5.5.1 Reference Resolution in Multilingual Dialogue Systems

In order to identify what a reference expression means (e.g. the pronoun 'it', the ordered expression 'the first one', etc.), the dialogue context must be stored in the dialogue manager. This module typically contains a *discourse modeller* sub-module that stores contextual information. This information can be used for several purposes, one of them is helping in the resolution of the referring expressions. The simplest way to resolve the anaphoric reference of a pronoun is to match the most recently mentioned entity that syntactically matches the pronoun. Using this strategy, in the following sample dialogue the 'ABC hotel' in turn S1 can be the referent for the pronoun 'it' in turn U1:

S1: The ABC hotel costs 10,000 Yen per night.
U1: O.K. I'd like to reserve it.

However, this simple reference resolution strategy causes a problem in MLDSs since the set of recently mentioned entities (also called *objects*) for each syntactic slot is different in each language. For example, in English a human can be referred by '*he*' or '*she*', while in French the corresponding pronoun '*il*' or '*elle*' can refer to a thing, such as a book or a house. This *recency* reference resolution strategy also fails in the following example:

S1: The ABC hotel has an in-door pool and an athletic room.
U1: O.K. I'd like to reserve it.

In this case, the recency strategy selects 'athletic room' in S1 as the referent for 'it' in U1. Some kind of centring mechanism must be used to deal with such a problem (Brennan et al. 1987; Grosz et al. 1995). The most salient object which is the first candidate of the referenced object is decided by the dialogue context. For example, in English the salience of an object is ordered by the following rule:

Topic > subject > object/object(indirect) > others

This ordering constraint is derived from syntactic information about the object. Since syntax is different across languages, such an ordering constraint is necessarily language-dependent. If the MLDS implements the interlingua architecture, such syntactic information could be assimilated because the interlingua is designed as a semantic representation independent of the syntactic features of each individual language. In order to use such syntactic salience ordering constraint for reference resolution, two possible methods are:

1. Preserve in the interlingua the syntactic features for each noun in the user's utterance.
2. Place the discourse level processing before the generation of the interlingua.

The first method violates the language independency principle of the interlingua and thus provokes maintenance problems. The second method forces discourse level processing in the NLP module, not only by referring dialogue context information but also by updating the dialogue context. This violates the system design principle of module independency, concretely between the NLP module and the dialogue manager.

5.5.2 Ambiguity of Speech Acts in Multilingual Dialogue Systems

The ambiguity of speech acts generally occurs in the processing of *indirect* speech. An example is the sentence 'Do you know the time?'. If a user asks this question to a dialogue system, it will simply answer 'Yes' or 'No'. However, this answer may not satisfy the user, who may be interested in knowing the current time. In order to give the appropriate answer, the system will need a plan recognition model to infer the user intention (Allen 1995). Although common in spontaneous human-to-human dialogues, this type of indirect speech does not seem to be frequent in human-to-system dialogues. However, the diversity of a user's utterances yields similar problems. One is shown in the following example.

S1: The ABC hotel costs 10,000 Yen per night.
U1: OK. Can I reserve a room?

If the 'ABC hotel' is the only hotel in the dialogue context, 'Yes, you can' would be an inappropriate response, since the system should take the user utterance as a request for reservation and then should start a subdialogue for this task. On the contrary, if there are several candidate hotels in the dialogue context, after searching for a hotel that meets some conditions, 'Yes, you can reserve a single room' can be an appropriate answer. This inference process needs a mechanism (e.g. implemented as a set of rules) that converts the surface expressions into illocutionary force types under the dialogue context. Since this set of rules tends to be some kind of idiomatic correspondence between surface expressions and illocutionary force types, it largely depends on each individual language (especially in pragmatics). Therefore, the rules should be applied before the semantic representation (interlingua) is generated, which would violate the module independency principle.

5.5.3 Differences in the Interactive Behaviour of Multilingual Dialogue Systems

A problem concerned with the interactive behaviour of MLDSs is the generation of the so-called *back-channelling* to provide information to the user about the system's current mode, e.g. listening or processing (not listening) mode. In order to set up a natural back-channelling feedback from the system, Cathcart et al. (2003) proposed a shallow model for the back-channelling generation that uses a pre-determined number of words, pause duration and POS (Part-of-Speech) trigrams. The number of words and POS trigrams appropriate for the back-channelling generation are language-dependent parameters. In addition, the algorithm to select the kind of back-channelling expression appropriate for a given situation (e.g. *mm-hmm*, *uh-hum*, etc.) is also language-dependent. Therefore, it is difficult to put the back-channelling generation mechanism into the dialogue management module, which forces language-dependent differences in the behaviour of this module. To address this problem, Fujie et al. (2004) presented a reactive approach based on using linguistic and non-linguistic information.

Another problem concerned with the interactive behaviour of MLDSs is the need for different behaviours depending on the user's barge-in point. If the system's prompt is too long, or the user is an expert using the dialogue system, the user will tend to interrupt the system's prompt to answer it. If the system's prompt is not a simple question but includes some information for the user, the system has to know whether such information is understood (or *grounded*) by the user. For this problem, the specification of SSML (Speech Synthesis Markup Language)[10] includes a mark element to indicate to the hosting environment (e.g. based on VoiceXML[11]) where the user's barge-in occurs. Using this mark element, the system can know what information has reached the user. For example, if the system's prompt consists of an information part and a question part, the following VoiceXML prompt with SSML markup can be used to identify where the user stopped the system prompt (Figure 5.6).

The problem is that if the content of the prompt element is made from interlingua (i.e. from the output of the dialogue manager), the mark elements have to be inserted according to

---

[10] http://www.w3.org/TR/speech-synthesis/
[11] VoiceXML will be studied in Chapter 6 and Appendix A.

```
<prompt>
        <mark name="info_start"/>
            information part
        <mark name="info_end"/>
            question part
        <mark name="question_end"/>
</prompt>
```

Figure 5.6   System prompt with barge-in markers

the structure of the message. Note that for some languages it may be preferable to place the information part first and then the question part, while for others languages the reverse order may be preferable. Moreover, the markers' semantic role and proper place are determined by task level information. Therefore, a NLG module that transforms interlingua into marked messages should be necessarily language and task-dependent.

5.6 Implications of Task Independency in the Dialogue Management

This section discusses the implications of task independency in dialogue management, presenting an initial classification of dialogue tasks into task classes (basically *slot-filling*, *database search* and *explanation*), and then focusing on the necessary modifications in these classes to achieve task independency.

5.6.1 Dialogue Task Classification

There are many application domains (*tasks*) for which dialogue has been implemented, such as telephone shopping, on-line trading, on-line banking, book order, route direction, tourist information, etc. These tasks can be classified into classes considering the direction of the information flow (Table 5.1) (Araki et al., 1999). The type of dialogue, subdialogue and interaction depends on the dialogue partner whose aim is to obtain information from the other dialogue partner.

5.6.1.1 Slot-filling Task Class

The *slot-filling* task class is the simplest one. Typical tasks in this class are telephone shopping, on-line banking, on-line trading, etc. The user aims to obtain information from the system and knows almost all the data necessary to carry out the task. Most (sub)dialogues

Table 5.1   Classification of dialogue tasks

Direction of information flow	Task class	Example task
User → System	Slot-filling	On-line shopping
User ↔ System	Database search	Book search
User ← System	Explanation	Route direction

in this task class are carried out to obtain values for the slots of a task frame. A sample dialogue taken from telephone shopping task can be as follows:

O: (Operator) Your telephone number please.
C: (Customer) 0262, 55, XXXX.
O: 0262, 55, XXXX. Is that right?
C: Right.
O: So, what is your favourite commodity?
C: A steel rack. Code number XC-6.

A dialogue system designed for this task class can basically be a slot-filler server side program. The user-system interaction is mostly system-directed. The task begins with an opening system's prompt to identify the service, continues with system prompts to collect values for the slots, typically one by one, and finishes when all the necessary slots are filled with the appropriate values.

5.6.1.2 Database Search Task Class

The *database search* class task has been implemented in many dialogue systems. Typical tasks in this class are ATIS, on-line book order, information retrieval, etc. The user aims to obtain information but only knows partial data in order to achieve the task. A sample dialogue taken from the literature retrieval task can be as follows:

C: (Client) I'd like to find out a paper about dialogue management.
S: (System) There are 2041 articles in the database. You must add additional information.
C: Then, add Japanese and reinforcement learning in keywords.
S: There are no articles. The added keywords may be too specific.

Dialogue systems for this task class typically interact with a database management system (DBMS) using e.g. SQL. The dialogue proceeds to narrow down the candidate records and the task is finished when the user has obtained the appropriate record(s).

5.6.1.3 Explanation Task Class

Typical tasks in this class are related with providing help to use software programs or devices (e.g. a printer). The user has a problem but does not have enough knowledge to solve it. Most (sub)dialogues in this task class are devoted to identifying the problem and giving explanations. A sample dialogue taken from the software assistance task can be as follows:

C: I cannot print out the handout for the presentation.
S: Please select the appropriate explanation by its number.
   1. print out handout in Impress.
   2. make presentation in Impress
   3. print out document in Writer
C: 1
S: Select from menu bar, file, print. Then a print dialogue box comes up.

A dialogue system for this task class typically uses a knowledge base and an explanation generator. If the system is implemented within a Web site, the knowledge to be explained is typically represented in XML format. If the explanation is provided through a GUI browser, a CSS (Cascading Style Sheet) is typically used for the rendering. The speech-based explanation requires some kind of rendering engine to divide the explanation into suitable pieces. The task begins identifying the information to be explained, and finishes when this information has been explained to the user.

5.6.1.4 More Complex Tasks

Needless to say, the three task classes explained above cannot cover all the tasks carried out by dialogue systems. However, these classes and their combination can cover a number of practical tasks with manageable vocabulary size, utterance patterns and (sub)dialogues. For example, the task concerned with 'trouble shooting on software installation' can be represented combining the *database search* and the *explanation task* classes, as shown in the following dialogue.

C: I have installed TeX and pLaTeX package but I cannot generate a postscript file.
S: What is your operating system?
C: Kyoto Linux.
S: Which version do you use?
C: Version 2.6.
S: Can you generate a dvi file?
C: Yes, I can. But I cannot generate a postscript file.
S: Did you install the dvi2ps package?
C: Yes.
[*** Database search phase has finished and Explanation phase begins. ***]
S: Then, it may be a font problem. Check if pk font files are installed.
C: How do I check it?

The first half of this dialogue is devoted to identifying the problem (FAQ *database search* task class) while the second is concerned with operational instructions (*explanation* task class).

5.6.2 Task Modification in Each Task Class

In this section we focus on the necessary modifications in these task classes in order to achieve the task independency.

5.6.2.1 System Architecture for Slot-Filling Task Class and Task Modification

In the *Slot-filling* task class, the dialogue management is rather simple compared to the other task classes. A simple dialogue system for this task class can be based on static VoiceXML files, using simple server-side dialogue management. The dialogue consists of an opening part, system-initiative for one-by-one slot filling subdialogues, confirmation subdialogues, report of results, and closing part. Each dialogue can be represented as an static VoiceXML file, in which the conditional branch of the dialogue management is based on the returned

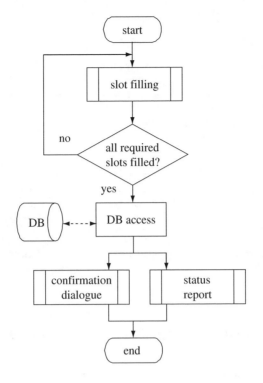

Figure 5.7   Dialogue flow for slot-filling task

values of application calls (e.g. result of requesting the reservation of a hotel room). Therefore, simple server-side programs (CGI or Servlet) can manage the dialogue in most cases.
The dialogue flow and required modules for this task class can be as shown in Figure 5.7.

Modifying this task class in order to achieve task independency requires defining a
sequence of slots, their data types (e.g. number, date, time, place name, etc.), writing the
corresponding prompts and grammars for each slot, and modifying the application APIs
according to the slot types and their data structures. For such modifications it is useful to use
a data-oriented architecture, or an Object-to-RelationalDB mapping such as in Hibernate,[12]
which can easily handle and convert between Java objects and database records.

5.6.2.2 System Architecture for Database Search Task Class and Task Modification

In the *Database search* task class, the central point of the dialogue management is how
to narrow down the number of possible retrieved records from the database, or how to
reformulate the user's query if no records are retrieved. Dialogue systems designed for this
task class typically use a database access module to transform the user input into database
queries (e.g. using SQL) and collect records from the database. The dialogue flow and
required modules for this task class can be as shown in Figure 5.8.

---

[12] http://www.hibernate.org/

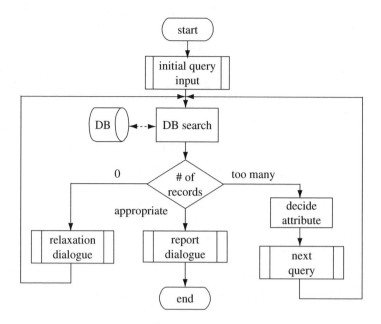

Figure 5.8   Dialogue flow for database search task

Modifying this task class in order to achieve task independency requires the following tasks:

- Prepare the grammar for the user's query input.
- Modify the database search operation according to the task data model.
- Modify the attribute decision rules called when the number of retrieved records exceeds a maximum value, as well as the relaxation dialogue in the opposite case.

Hamerich et al. (2004a, 2004b) proposed a semi-automatic, three-layered platform to generate dialogue systems for this task class. In this platform, the higher layer defines the data model and the data connection function; the middle layer defines the interaction for retrieving data (in a language- and modality-independent way); and the lower layer specifies the interaction patterns using VoiceXML and XHTML.

5.6.2.3 System Architecture for Explanation Task Class and Task Modification

In the *Explanation* task class, the critical point is not in the dialogue management but in the recognition and analysis of the user query. The reason is that these queries can be very diverse, often including phenomena such as fillers, false starts, hesitation, etc., caused by the user's lack of knowledge about the task. Therefore, the first user's query must be recognised using an ASR statistical language model designed for the target domain. In addition, an appropriate query analyser must be used to extract the important content words and construct the query, which typically requires using a large-scale knowledge base. The dialogue flow and required modules for this task class can be as shown in Figure 5.9.

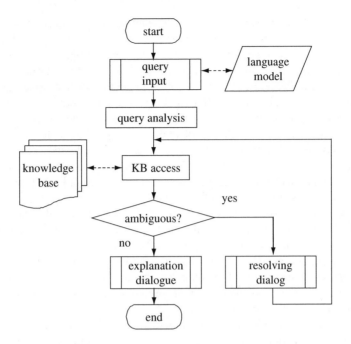

Figure 5.9   Dialogue flow for explanation task

Modifying this task class to achieve task independence requires the following:

- Using a dialogue corpus containing user queries in the target domain in order to build the ASR language model. Among others, Gao et al. (2005) discusses methods to enhance the portability of language models, such as interpolating domain-specific language models with other language models, and improving the system robustness and accuracy with limited in-domain resources.
- Set up the query analyser. If the corpus of user queries is accompanied by the answers taken from the knowledge base, this analyser can be constructed statistically.
- Set up two generation modules: one to generate explanation (sub)dialogues, and the other to generate ambiguity resolution (sub)dialogues. If the knowledge base is represented as structured documents (e.g. XML), the generation modules must use appropriate NLG techniques to build fluent natural language expressions required for the posterior speech synthesis. On the contrary, if the knowledge base is represented directly in natural language, the generation modules can use information extraction (IE) techniques, since this knowledge base tends to include long explanations already expressed in natural language (e.g. software manuals).

5.7 Summary

In this chapter we first discussed the annotation of spoken dialogue corpora at the morphosyntactic, syntactic, prosodic and pragmatic levels, and then commented on some research efforts to stimulate the development of guidelines and means to share multimodal dialogue corpora. We also discussed some approaches for annotation of speech, gestures and other modalities. The second section of the chapter addressed dialogue modelling, mentioning some dialogue theories and focusing on the state-transition and plan-based dialogue models. The third section discussed dialogue management, describing different interaction strategies (user-directed, system-directed and mixed) and commenting on the selection of the most appropriate one. Confirmation strategies (explicit and implicit) were outlined and the use of confidence scores and adaptive methods to optimise these strategies were explained. The fourth section addressed the implications of multimodality in the dialogue management. It initially discussed the implications in terms of interaction complexity, commenting on why the complexity is in general greater for MMDSs than for SDSs. Next, it focused on the implications in the confirmation strategy, which is also more complex but more efficient for MMDSs than for SDSs. Implications of the so-called *social* and *emotional* dialogue have also been addressed. As has been noted, the social dialogue requires the setting up of more sophisticated dialogue strategies than in the case of pure, task-oriented systems, while the emotional dialogue requires the processing of *emotions*, which are considered as additional input/output interaction modalities. Next, the implications of multimodality in terms of contextual information, which is crucial to obtain the user intention and discard wrong hypotheses, were discussed. Unification and overlay operations were explained to deal with this kind of information. Then, the effects of multimodality on the user references, which are much more complex in the case of MMDSs, as can be cross-media and cross-user/system, were presented. The fourth section finally discussed the implications in terms of response generation, which is also more complex as MMDSs must decide the output to be produced through each modality and coordinate the output across the modalities. The fifth section addressed the implications of multilinguality in dialogue management, discussing the resolution of references, the ambiguity of speech acts, and the differences in the interactive behaviour of MLDSs. The final section of the chapter discussed the implications of task independency in dialogue management. It presented a generic classification of tasks typically performed by dialogue systems into task classes (basically slot-filling, database search and explanation) and outlined the necessary modifications in these classes to achieve task independency.

5.8 Further Reading

Dialogue Annotation

- Chapter 1 of Gibbon et al. (2000).
- http://www.cogsci.ed.ac.uk/~amyi/mate/report.html

Dialogue Modelling

- Part I of Taylor et al. (2000).
- Chapter 3 of McTear (2004).

Dialogue Management

- Part I of Taylor et al. (2000).
- Chapter 5 of McTear (2004).

6

Development Tools

6.1 Tools for Spoken and Multilingual Dialogue Systems

This section initially addresses tools to develop some modules of SDSs, concretely HTK for the ASR module, CMU-Cambridge Statistical Language Modelling Toolkit for the language modelling, and Festival and gtalk for speech synthesis.

6.1.1 Tools to Develop System Modules

6.1.1.1 Automatic Speech Recognition

The recent progress in computing power, storage capabilities and availability of large amounts of machine-readable text and speech corpora, has promoted important advances in statistical speech recognition methods and tools. HTK[1] (Hidden Markov Model ToolKit), developed by Cambridge University, is a toolkit that enables to use Hidden Markov Models (HMMs) for statistical modelling, especially for speech recognition. The toolkit consists of a set of libraries and tools written in C code, which are platform-independent.

HTK offers a set of commands for recording data, and for initialisation, training and evaluation of HMMs. In order to learn to use the toolkit, it is important to understand the basic steps to build HMMs and know the input/output files involved in each step. For simplicity, we describe in this section the construction of a word recogniser using the toolkit. The basic construction steps are the following (see Figure 6.1):

1. Speech data recording
2. Feature extraction
3. Definition of model parameters
4. Training
5. Evaluation

---

[1] http://htk.eng.cam.ac.uk/

Spoken, Multilingual and Multimodal Dialogue Systems: Development and Assessment Ramón López-Cózar Delgado and Masahiro Araki © 2005 John Wiley & Sons, Ltd

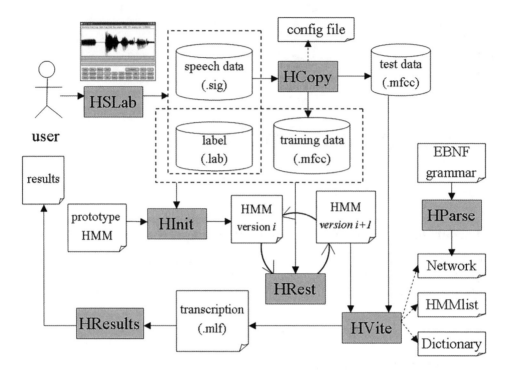

Figure 6.1 Using HTK for speech recognition

In the speech data recording step, the **HSLab** tool can be used for recording and labelling. The default sampling rate is 16 kHz and the default file format is a specific HTK format (.sig file). Other speech recording tools which output .wav files can be also used. The labelling is an operation that divides the speech signal into units. Each unit corresponds to one HMM.

A sample labelling result is shown in Figure 6.2 (each line contains a start segment, a final segment and a label name). The second step is the feature extraction from the speech data, which can be carried out using the **HCopy** command. This command typically needs two parameters: a configuration file that specifies the parameters of the acoustical analysis and a script file that specifies the path for the input speech data files and the output feature files. A sample configuration file is shown in Figure 6.3. The script file is a sequence of pairs of input files (.sig) and output files (.mfcc). The output data must be divided into two groups (one for training and the other for testing).

```
131250 2890000  Ns
2933750 7226250 one
7313750 13094375 Ns
13138125 16685625 two
16729375 21065625 Ns
21109375 24919375 three
24875625 29911875 Ns
```

Figure 6.2 Output format of speech labelling (Ns: noise)

```
# Coding parameters
TARGETKIND = MFCC_0        # coefficients to use (0 means energy)
TARGETRATE = 100000.0      # 10 ms frame period
WINDOWSIZE = 250000.0      # 25 ms window size
USEHAMMING = T             # use of Hamming function
NUMCHANS = 26              # number of filterbank channels
CEPLIFTER = 22             # length of cepstral liftering
NUMCEPS = 12               # number of MFCC coefficients
```

Figure 6.3 Example of configuration file for the **HCopy** command

The third step is the definition of the model parameters and the making of the initial HMM for each recognition unit, such as phoneme, word, etc. The definition file (Figure 6.4) can be created using a text editor. The first line is the header of the file and specifies the vector size and type of features. The second line identifies the class label for which the HMM is defined. The structure of the HMM is defined within the tags `<BeginHMM>` and `<EndHMM>`. The first element of this structure is the number of states for the HMM, including the 'non-emitting' states (i.e. the first and the last states). Thus, for example, if the HMM has 3 active states (i.e. 5 states in total), this section must define the parameters for states 2, 3 and 4. The initial mean vector and initial variances of each dimension for each state are defined after the `<State>` tag. The last section defines the matrix of initial transition probabilities after the `<TransP>` tag.

After defining the prototype HMM, the initial parameters for the training step are calculated by the **HInit** command, which estimates the parameters using the Viterbi alignment. Another initial parameter estimation command is **HCompv**, which gives a flat initialisation to each model, i.e. each state of the HMM is given the same mean and variance vectors, which are computed from all the training data.

```
~o <VecSize> 13 <MFCC_0>
-h "one"
<BeginHMM>
  <NumStates> 5
  <State> 2
    <Mean> 13
        0.0 0.0 0.0 0.0 0.0 0.0 0.0 0.0 0.0 0.0 0.0 0.0 0.0
  <Variance> 13
        1.0 1.0 1.0 1.0 1.0 1.0 1.0 1.0 1.0 1.0 1.0 1.0 1.0
  <State> 3
        ...
  <State> 4
        ...
  <TransP> 5
        0.0 1.0 0.0 0.0 0.0
        0.0 0.5 0.5 0.0 0.0
        0.0 0.0 0.5 0.5 0.0
        0.0 0.0 0.0 0.5 0.5
        0.0 0.0 0.0 0.0 0.0
<EndHMM>
```

Figure 6.4 Definition of HMM structure and parameters

The fourth step is the training, which is carried out by the **HRest** command making an iteration of the Baum-Welch re-estimation algorithm. The input to this command is the HMM defined in the previous step or trained in the previous iteration. The convergence in the training can be examined by the *change* measure which is displayed during the each iteration.

The last step is the evaluation of the HMM, which can be done using the **HVite** command. If the test is concerned with word recognition, the candidate words are defined in the task grammar file, which in this case simply lists possible words. The dictionary file contains all the words of a particular task that can be recognised by the system. If the test is concerned with recognising sequences of words, the construction (grammar) rules are defined in the task grammar file by EBNF (Extended BNF) format. In this case, the task grammar file contains all the possible sequences of words (sentences) that can be recognised. The task grammar file is converted into a word network file by the **HParse** command.

With this information as parameters, the **HVite** command performs the Viterbi algorithm to carry out the recognition and provide the transcription (.mlf). The recognition error rate can be calculated by the **HResults** command.

6.1.1.2 Language Modelling

The CMU-Cambridge Statistical Language Modeling Toolkit[2] is widely used to make ASR statistical language models. This toolkit consists of a set of commands for statistical processing that eases the rapid construction of arbitrary N-gram models (e.g. unigrams, bigrams, trigrams, etc.) with four types of built-in back-off algorithms (Good Turing, Witten Bell, Absolute and Linear discounting). The basic procedure to build a N-gram language model is divided into the following three steps (see Figure 6.5):

1. Making of the vocabulary
2. Making of the N-gram model
3. Evaluation.

In order to make the language model, large amounts of text data (corpus) must be prepared in advance. The words in the corpus are assumed to be separated by orthographic markers (e.g. white spaces). Therefore, languages that have no explicit word separation markers, e.g. Japanese, must be segmented by a morphological analysis in advance. As optional information, sentence and paragraph boundary markers (e.g. <s> and </s>) can be inserted in the text corpus. For the purpose of evaluation, the corpus should be divided into two groups: one for training and the other for testing.

The first step to create a language model is to make the vocabulary for the target language. The **text2wfreq** command can be used to calculate the word frequency. The output file of this command (.wfreq) contains the different words in the corpus along with the number of their occurrences. The command reports about possible formatting problems in the corpus (e.g. inserted parentheses). In order to limit the size of the model, the vocabulary must be limited by a certain criteria. The **wfreq2vocab** command uses the word unigram file (.wfreq)

---

[2] http://mi.eng.cam.ac.uk/~prc14/toolkit.html

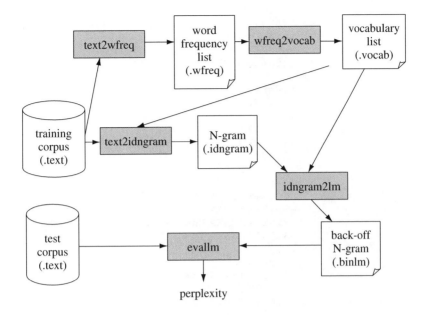

Figure 6.5 Definition of HMM structure and parameters

and selects the top N words by their frequency, or the words that appear more than M times, creating a vocabulary file (.vocab) that is used in the next step.

The second step is to make the N-gram model from the training corpus. The words defined in the vocabulary file appear in the N-gram model, while others are treated as out-of-vocabulary (OOV) words. Using the vocabulary file and the training corpus, the **text2idngram** command generates an id N-gram file (.idngram), which contains a numerically sorted list of N-tuples of numbers, corresponding to the mapping of the word N-grams relative to the vocabulary.

The OOV words are mapped to the number 0. The .idngram file is used by the **idngram2lm** command to make the statistical language model using a discounting algorithm specified as a parameter. The output can be either a binary file (.binlm) or a text file (arpa format). Figure 6.6 shows an example of file in arpa format.

The final step is the evaluation of the language model, which can be carried out by the **evallm** command using the test corpus. In general, statistical language models are evaluated in terms of word perplexity. A sample evaluation result is shown in Figure 6.7.

6.1.1.3 Speech Synthesis

Festival

The Festival Speech Synthesis System[3], developed by The Centre for Speech Technology Research (CSTR) at University of Edinburgh, is a general purpose speech synthesiser that

---

[3] http://www.cstr.ed.ac.uk/projects/festival/

```
\data\
ngram 1=192
ngram 2=1490
ngram 3=3408

\1-grams:
-0.4562 <UNK> -0.5037
-1.3271 </s>  -2.5304
-1.3259 <s>   -1.2636
-2.9742 ASR   -0.3053
-2.9742 Among -0.7948
...
\2-grams:
-0.4448 <UNK> <UNK> -0.0159
-1.1811 <UNK> </s> 0.3109
-3.0274 <UNK> English 0.6573
-3.6542 <UNK> In -0.1537
...
\3-grams:
-0.7203 field in which
-0.4192 field of dialogue
-0.4192 field that <UNK>
...
```

Figure 6.6   Example of text file (arpa format)

```
$ evallm.exe-arpa data.arpa
Reading in language model from file data.arpa
Reading in a 3-gram language model.
Number of 1-grams=192.
Number of 2-grams=1490.
Number of 3-grams=3408.
Reading unigrams...
Reading 2-grams...
Reading 3-grams...
Done.

evallm : perplexity-text test.text
Computing perplexity of the language model with respect
  to the text test.text
Perplexity = 39.64, Entropy = 5.31 bits
Computation based on 11719 words.
Number of 3-grams hit=7206 (61.49%)
Number of 2-grams hit=2895 (24.70%)
Number of 1-grams hit=1618 (13.81%)
9191 OOVs (43.96%) and 0 context cues were removed from the
  calculation.
evallm : quit
evallm : Done.
```

Figure 6.7   Evaluation of the language model

can be used from a shell command via a Scheme command interpreter and from C++ and Java programming languages. It consists of externally configurable language independent modules, such as phone sets, lexicons, letter-to-sound rules, tokenising, part of speech tagging, and intonation/duration. It is a diphone-based waveform synthesiser whose diphone database can be replaced by other diphone databases such as MBROLA[4]. The current version works for English (British and American), Welsh and Spanish.

The shell command mode is invoked by the "festival" command. It shows the version and copyright information and waits for a user's command.

```
%festival
Festival Speech Synthesis System 1.95:beta July 2004
Copyright (c) University of Edinburgh, 1996-2004. All rights
reserved.
For details type '(festival_warranty)'
festival>
```

This shell is a kind of Scheme programming language. For example, the simplest speak command is the following:

```
festival> (SayText "My name is Festival")
```

To get a waveform from the text, the synthesis process can be done stepwise. The first step is to assign the text to a variable.

```
festival> (set! utt1 (Utterance Text "My name is festival"))
```

Then, the following command generates a waveform by identifying tokens, converting letters to phoneme sequence, deciding their duration and adding an intonation pattern.

```
festival> (utt.synth utt1)
```

The waveform can be saved to a file or played back via the audio device by using the following commands:

```
festival> (utt.save.wave utt1 "utt1.wav" "wav")
festival> (utt.play utt1)
```

The input for the system can be a either plain text or a markup language called Sable[5], which has several elements to control the synthesised voice, such as speech rate, pitch, emphasis and volume.

---

[4] http://tcts.fpms.ac.be/synthesis/mbrola.html
[5] http://www.bell-labs.com/project/tts/sable.html

Flite

The Flite[6] system, developed by CMU (Carnegie Mellon University, USA) as a derivation of Festival, is a light-weight and fast run-time speech synthesis engine. It was designed for embedded systems like PDAs and parallel execution in backend speech application servers. It was written in ANSI C, and designed to be portable to almost any platform with scalable voice size. The API is so simple that the C code for using the this system can be short and concise. Examples of public functions are: **flite_init** (for initialisation), **flite_text_to_wave** (to return a waveform), and **flite_text_to_speech** (to send the synthesised text to the audio device). The Java version of Flite is FreeTTS[7], developed by the Speech Integration Group of Sun Microsystems Laboratories. Since JSAPI (Java Speech API) is implemented on this synthesiser, the system is platform-independent provided that the Java2 SDK (above version 1.4) is installed.

MBROLA

The MBROLA project[4] was initiated at the TCTS Lab of the Faculté Polytechnique de Mons (Belgium). The aim of this project is to obtain a set of speech synthesisers for as many languages as possible, and make them free available for researchers and non-commercial applications. As a result, MBROLA is a speech synthesiser based on diphone concatenation that has as input a list of phonemes with prosodic information (duration and pitch pattern) and produces speech output at the frequency of the diphone database being used. Thus, MBROLA is not TTS (Text-To-Speech) system since it does not have a linguistic analyser to process raw text input.

Galatea talk

Galatea talk (or 'gtalk') is a speech synthesiser for Japanese that was released by the Interactive Speech Technology Consortium (ISTC[8]). It is as a module of the Garatea toolkit, designed for building anthropomorphic spoken dialogue agents. The basic architecture of this speech synthesiser is a state machine model. Thus, it can be controlled by commands and parameter settings through socket communication (Windows version) or standard I/O (Unix version).

In the linguistic analysis, a Japanese morphological analyser is used, ChaSen[9], that provides a phoneme sequence and an accent pattern for each word. Accent pattern modification rules for concatenated words are also applied. The wave form generation relies on a HMM-based method that allows to use speaker adaptation techniques employed for HMM-based speech recognition.

Another remarkable feature of this system is the exact measurement of the duration time (in ms) for each phoneme (Figure 6.8 shows the duration within braces, e.g. k[100] means the duration of the phoneme /k/ is 100 ms). This information is useful to identify the break point when the user's barge-in occurs (i.e. the system can grasp which part of the text has been delivered to the user).

[6] http://www.speech.cs.cmu.edu/flite/
[7] http://freetts.sourceforge.net/docs/index.php
[8] http://www.astem.or.jp/istc/index_e.html
[9] http://chasen.naist.jp/hiki/ChaSen/

```
rep Run=LIVE
rep Speak.stat=PROCESSING
rep Text.text=これは音声合成のテストです
rep Speak.text=これは音声合成のテストです
rep Text.pho=sil[340] K[100] o[55] r[30] e[100] w[75] a[80] o[70]
   N[75] s[110] e[30] e[65] g[45] o[60] o[65] s[90] e[30] e[85] n[40] o[80]
   t[80] e[65] s[90] U[25] t[60] o[65] d[50] e[65] s[165] U[25] sil[340]
rep Speak.pho = sil[340] k[100] o[55] r[30] e[100] w[75] a[80]
   o[70] N[75] s[110] e[30] e[65] g[45] o[65] s[90] e[30] e[85] n[40] o[80]
   t[80] e[65] s[90] U[25] t[60] o[65] d[50] e[65] s[165] U[25] sil[340]
rep Text.dur=2655
rep Speak.dur=2655
rep Speak.stat=READY
rep Speak.stat=SPEAKING
rep Run = EXIT
rep Speak.len=2655
rep Speak.utt=sil[340] k[100] o[55] r[30] e[100] w[75] a[80] o[70]
   N[75] s[110] e[30] e[65] g[45] o[60] o[65] s[90] e[30] e[85] n[40] o[80]
   t[80] e[65]s[90] U[25] t[60] o[65] d[50] e[65] s[165] U[25] sil[340]
rep Speak.stat=IDLE
```

Figure 6.8   Execution log of the Galatea talk

6.1.2 Web-Oriented Standards and Tools for Spoken Dialogue Systems

In the 1980s and the 1990s the developers of SDSs had to set up their systems using general-purpose programming languages and, in general, needed to have a deep knowledge about the diverse technologies required for the setting up. The implementation was facilitated by the availability in the 1990s of API for ASR and TTS modules, as well as tools to set up some components (e.g. HTK). Recently, as a result of the efforts made by the Voice Browser Activity of the World Wide Web Consortium (W3C), new XML-languages have been developed to ease the implementation of SDSs by applying Web technologies. The purpose of this Activity is to enable the user to access services from the telephone, via a combination of speech and DTMF. The Activity has developed a suite of specifications for standardising the implementation of SDSs, which are shown in Figure 6.9 (the rectangular boxes represent components of the SDS and the ellipses represent specifications related to each component).

The central specification of the Activity is VoiceXML (Voice eXtensible Markup Language) which describes the user-system dialogue, the processing of the user input and the communication with a document server.

SRGS (Speech Recognition Grammar Specification) is a standardised grammar format, either in Augment BNF form or XML form, that specifies the words and sequences of words than can be recognised. SSML (Speech Synthesis Markup Language) is a standardised language to assist the generation of speech in Web and other applications. The language gives users the chance to enhance the quality of the synthesised speech by using parameters such as volume, pitch, rate, etc.

The standard TTS modules employed by SDSs may have problems in synthesising proper nouns (e.g. persons' names, place names, etc.) or other unusual words, as they may not be in the platform's built-in lexicons. In this case, the PLS (Pronunciation Lexicon Specification) can be used to supply high quality pronunciation in a platform-independent way.

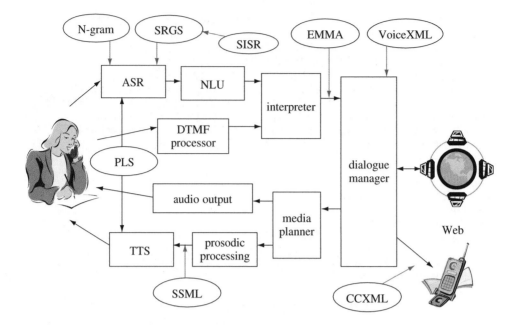

Figure 6.9   Overview of the W3C Voice Browser Activity

CCXML (Call Control XML) provides the functionally that VoiceXML lacks in terms of
call control. It defines how calls are placed, answered, transferred, etc. It is a language that
must be used in combination with VoiceXML.

SISR (Semantic Interpretation for Speech Recognition) describes annotation rules to use
with SRGS for extracting the semantics from ASR results. The Activity defines a target data
format called EMMA (Extensible MultiModal Annotation markup language), to represent
the information contained in the user input, regardless of the interaction mode. It is specified
as the information exchange format between modules in the input and output components. It
includes semantic representations by the annotations of several features made by input and
output modules, such as timestamps, confidence scores, grouping information, etc. The goal
is that SISR generates data that can be integrated in EMMA.

The stochastic language models (N-Gram) specification defines the syntax for representing N-Gram (Markovian) stochastic grammars, which support large vocabulary and open
vocabulary applications. Moreover, these grammars can be used to represent concepts or
semantics.

6.1.2.1 VoiceXML

VoiceXML is an XML-based language that is considered the standard language to access
Web contents using speech or DTMF. This language facilitates the development of SDSs
relying on the Internet client/server architecture. Using this standard, the system can interact
with the user via synthesised speech or pre-recorded messages, while the user can interact

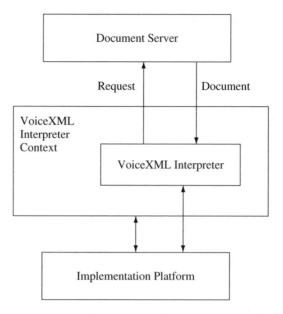

with the system using speech or DTMF. At the time of writing, there are two versions of VoiceXML: v1.0 made public in 2000[10] and v2.0 made public in 2001.[11] Version 1.0 was developed by the VoiceXML Forum[12] and submitted to the World Wide Web Consortium[13] (W3C), which published it as a Note. Version 2.0 was developed through the W3C recommendations track process. Figure 6.10 shows the base architecture for this language, which contains four main modules: VoiceXML Interpreter, Document Server, VoiceXML Interpreter Context and Implementation Platform[14].

The VoiceXML Interpreter resides in the client application. It executes the logic of the application, plays prompts to obtain data from the user which are necessary to provide a service, processes the user requests, and searches for the information requested by the user in Web servers or local databases. The Document Server is a Web server that processes the requests sent by the VoiceXML Interpreter and returns VoiceXML documents. The VoiceXML Interpreter Context processes the VoiceXML documents, answers user calls and supervises the user input, generating predefined messages in response to some events, for

---

[10] http://www.w3.org/TR/2000/NOTE-voicexml-20000505/

[11] http://www.w3.org/TR/voicexml20/

[12] http://www.voicexml.org/

[13] http://www.w3.org

[14] This document has been reviewed by W3C Members and other interested parties, and it has been endorsed by the Director as a W3C Recommendation. W3C's role in making the Recommendation is to draw attention to the specification and to promote its widespread deployment. This enhances the functionality and interoperability of the Web. This specification is part of the W3C Speech Interface Framework and has been developed within the W3C Voice Browser Activity by participants in the Voice Browser Working Group.

example, when the user asks for help, does not answer a system's prompt, etc. The last component of the VoiceXML architecture is the Implementation Platform, which contains the telephone hardware and other resources. This module activates events in response to user actions (e.g. keys pressed) and in other circumstances (e.g. a timer timeout). To interact with a VoiceXML application, the user places a telephone call which crosses the telephone network and arrives at the so-called *voice site*, which is a computer that behaves as the interface between the telephone network and the Internet. It contains all the modules mentioned above but the Document Server, which may be in another computer.

A VoiceXML application is a set of VoiceXML documents that share a root document. When the user interacts with a document of such an application, its root document is always stored in the system memory and is removed when the interpreter makes a transition to a document that belongs to another application. A grammar determines the vocabulary and all the sentences allowed in the corresponding dialogue state.

The grammars and variables of the root document are always active, independently of the application document in which the user is interacting at a given moment. A dialogue state can have one or more grammars associated. *Links* are used to specify transitions to other parts of the same document, other documents of the same application, or documents of other applications.

A VoiceXML document is a set of high-level elements called *dialogues*. There are two types of dialogues: *forms* and *menus*. The former are used to obtain data from the user using *fields*. A *form* has an optional identifier (its name), can define variables and may include control elements, event handlers and 'filled' blocks, which contains code that is executed when determined fields are filled from the user input. *Menus* are used to give options to the user and make transitions to other states depending on his input. In addition to dialogues, in VoiceXML there are also *subdialogues*, which are similar to functions of high-level programming languages, given that when a subdialogue finishes, the interpreter makes a transition back to the calling point.

Figure 6.11 shows a simple application written in VoiceXML 1.0 as well as a possible computer-human dialogue. The application only contains a *form* inside of which there is a *field* named '`drink`' whose value must be obtained from the user during the dialogue. The *prompt* identifies the message the system must generate via TTS in order to obtain the value for the field. The *grammar* identifies the file ('`drink.gram`') that contains the recognisable vocabulary and sentence types for the ASR. Once the data is obtained from the sentence uttered by the user, the *submit* element sends it ('`coffee`', '`tea`', '`milk`' or '`nothing`') to a Web server where it is processed.

Appendix A presents a basic, introductory tutorial to VoiceXML. We recommend the reader studies the following issues (among others), either on the W3C Web page or in some book on VoiceXML (see Further reading section):

- Fundamentals.
- Form Interpretation Algorithm (FIA) and its fine-grain control (i.e. how to alter the way in which form items are visited).
- Form items and guard conditions.
- Interaction strategies (system-directed and mixed).

- Links' definition and scope.
- Declaration, assignation and scope of variables.
- Definition, types and scope of grammars.
- Event handling (generation and management).
- Resource fetching and attributes (caching, fetchtimeout and fetchhint).
- Message recording.
- Call transfer (bridging and blind transfer).
- Execution of actions when form fields are filled in.
- Field names' hidden variables (e.g. confidence, input mode, etc.).
- Platform properties.
- Executable content.
- Conditional logic (if-then-else).
- Prompt generation (*tapered* prompts and *reprompt*).
- Transitions (intra- and inter-documents).
- Submission of variables to document servers.
- Scripts (server-side code to carry out a specific action).
- Dynamic VoiceXML (use of Web technologies such as JSP, ASP, Perl, ColdFusion, etc. to generate new VoiceXML document at run time).

```
<?xml version="1.0"?>
<vxml version="1.0"?>
  <form">
    <field name="drink">
      <prompt>Would you like coffee, tea, milk, or nothing?</prompt>
      <grammar src="drink.gram" type="application/x-jsgf"/>
    </field>
    <block>
      <submit next="http://WWW.drink.example/drink2.asp"/>
    </block>
  </form>
</vxml>

C (computer): Would you like coffee, tea, milk, or nothing?
H (human): Orange juice
C: I did not understand what you said
C: Would you like coffee, tea, milk or nothing?
H: Tea
C: (contines in document drink2.asp)
```

Figure 6.11 Simple VoiceXML application and computer-human dialogue © Note 5 May 2000,[15] W3C (MIT, European Research Consortium for Informatics and Mathematics, Keio University).

---

[15] This document is a submission to the World Wide Web Consortium from the VoiceXML Forum. This document is a Note made available by the W3C for discussion only. This work does not imply endorsement by, or the consensus of the W3C membership, nor that W3C has, is, or will be allocating any resources to the issues addressed by the Note. This document is a work in progress and may be updated, replaced, or rendered obsolete by other documents at any time.

6.1.2.2 CCXML

CCXML (Call Control XML)[16] is another XML-based language that is also considered a standard developed by the W3C. It provides the functionally that VoiceXML lacks in terms of call control. Major features of this language are the handling of asynchronous events, multi-conference, multiple calls and placing of outgoing calls. It was designed to be used with VoiceXML, not separately. The first version was made public in 2002. In this section we give a short introduction to the use of some tags in order to provide a general overview of an application written in this language, as well as the main differences in comparison to VoiceXML. Figure 6.12 shows a simple CCXML application. It begins with the tags indicating the versions of XML and CCXML and ends with the `</ccxml>` tag. As can be observed, variables are defined very similarly as to VoiceXML, using the tag `<var name="..." expr="..."/>`. An initial difference between both languages is concerned with the use of prompts, since CCXML does not generate prompts, i.e. it does not use the tag `<prompt>`. To track the behaviour of CCXML applications it is a common practice to generate debug messages with the tag `<log expr="..."/>`, as can be observed in Figure 6.12. Running this application, the variable MyVariable is defined with the value "This is a CCXML Variable", and thus the generated log message is "'Hello World: I just made a variable: This is a CCXML Variable". Finally, the `<exit/>` tag is used to stop the execution and its thread of control.

CCXML works like a *state machine* in which several types of event change the current 'state'. To deal with this state machine, a variable is associated with a piece of code that processes the events, defined by the `<eventprocessor>` ... `</eventhandler>` tags. The variable is called the 'statevariable'. Figure 6.12 shows how simple it is to create a state machine: a variable named 'state0' is defined and initialised e.g. with

```
<?xml version="1.0" encoding="UTF-8"?>
<ccxml version="1.0">
  <var name="state0" expr="'init'"/>
  <eventprocessor state variable="state0">
    <transition event="conection.alerting" name="evt">
      <var name="MyVariable" expr="'This is a CCXML Variable'"/>
      <log expr="'Hello World. I Just made a variable:'+ MyVariable"/>
      <log expr="'let hang up on this incoming call as this is an example.'"/>
      <exit/>
    </transition>
  </eventhandler>
</ccxml>
```

Figure 6.12 Simple CCXML application © W3C Working Draft 30 April 2000,[17] W3C (MIT, European Research Consortium for Informatics and Mathematics, Keio University). All rights reserved

---

[16] http://www.w3.org/TR/2002/WD-ccxml-20021011/

[17] This specification describes the Call Control XML (CCXML) markup language that is designed to provide telephony call control support for VoiceXML or other dialog systems. This document has been produced as part of the W3C Voice Browser Activity, following the procedures set out for the W3C Process. The authors of this document are members of the Voice Browser Working Group.

the value 'init'. The <eventprocessor statevariable="state0"> is used to create the event processor for the application and the 'state0' variable is associated with it using the statevariable attribute. The state machine is now created and the variable 'state0' is its 'statevariable'.

Understanding the event processing is a basic issue in CCXML. Since events provoke transitions in the state machine, <transition> tags are used in the event processing section to catch the different event types and decide the actions to be carried out in response. Events are thrown asynchronously and are handled sequentially following the order of the <transition> tags in the event processor of the document. Some event types are the following: connection.alerting (event thrown when an incoming telephone call is detected), connection.connected (event thrown when the telephone call is being attended), connection.failed (event thrown when a telephone call cannot be established), etc. For example, Figure 6.13 shows a sample application to accept or reject a call depending on the telephone number of the calling party.

Up to now nothing has been said about the dialogue with the user, either using speech or DTMF. The reason is that CCXML is not designed for that purpose, but only to control telephone line events. For the dialogue with the user, a CCXML application makes a call to a VoiceXML document that handles the dialogue with the user as discussed in the previous section. This type of application is called *hybrid* since combines CCXML and VoiceXML execution. When the VoiceXML application ends, the flow of control is returned to the calling CCXML application. In the same way as in function calls, variables can be passed from CCXML to VoiceXML and returned from the latter to the former. The call to the VoiceXML

```
<?xml version="1.0" encoding="UTF-8"?>
<ccxml version="1.0">
 <eventprocessor>
   <transition event="conection.alerting" name="evt">
    <log expr="'The phone number the user called is'+evt.connection.
      remote+'.'"/>
    <if cond="evt. connection.remote == '8315551234'">
     <log expr="' Go away! we do not want to answer the phone."/>
     <reject/>
    <else/>
     <log expr='"we like you! we are going to answer the call.'"/>
     <accept/>
    </if>
   </transition>
   <transition event="conection.connected">
   <log expr="'Call has been disconnected. we should now end the CCXML
     session."/>
    <exit/>
   </transition>
 </eventhandler>
</ccxml>
```

Figure 6.13 Simple CCXML application to accept or reject an incoming call © W3C Working Draft 30 April 2000, W3C (MIT, European Research Consortium for Informatics and Mathematics, Keio University). All rights reserved

application is carried out using the CCXML tags `<dialogstart src="..."/>` and `<dialogterminate ... />`. For example, Figure 6.14 shows a sample CCXML application that calls a v2.0 VoiceXML application named 'gimme.vxml' (Figure 6.15) which returns a field named "input".

We recommend that the reader studies the following issues (among others), either at the W3C Web page or in some book on CCXML:

- Fundamentals.
- Basic processing of events.
- CCXML as a state machine.
- Call connection and disconnection.
- Error management.
- Combination of VoiceXML and CCXML: hybrid applications.
- Creation and destruction of a multiconference.

```xml
<?xml version="1.0" encoding="UTF-8"?>
<ccxml version="1.0">
 <!-- Lets declare our state var -->
 <var name="state0" expr="'init'"/>
 <eventprocessor statevariable="state0">
   <!-- Process the incoming call>
   <transition state="'init'" event="conection.alerting">
     <accept/>
   </transition>
   <!-- Call has been answered -->
   <transition state="'init'" event="connection.connected" name="evt">
    <log expr="'Houston, we have liftoff.'"/>
    <dialogstart src="'gimme.vxml'"/>
    <assign name="state0" expr="dialogActive'"/>
   </transition>
   <!-- Process the incoming call -->
   <transition state="'dialogActive'" event="dialog.exit" name="evt">
    <log expr="'Houston, the dialog returned ['+evt.values.input +']'"/>
    <exit/>
   </transition>
   <!-- Caller hung up. Lets just go on and end the session -->
   <transition event="connection.disconnected" name="evt">
    <exit/>
   <transition/>
   <!-- Something went wrong. Lets go on and log some info and end the
     call -->
   </transition event="error.**name="evt">
    <log expr="'Houston, we have a problem: (' + evt.reason + ')'"/>
    <exit/>
   </transition>
 </eventhandler>
</ccxml>
```

Figure 6.14 Simple CCXML application that calls a VoiceXML application © W3C Working Draft 30 April 2000, W3C (MIT, European Research Consortium for Informatics and Mathematics, Keio University). All rights reserved

```
<?xml version="1.0">
<vxml xmlns="http://WWW.W3.org/2001/vxml" version-"2.0">
 <form id="Form">
  <Field name="input" type="digits">
   <prompt>
    please say some numbers ...
   </prompt>
   <filled>
    <exit namelist="input"/>
   </filled>
  </field>
 </form>
</vxml>
```

Figure 6.15 Simple VoiceXML application called from the CCXML application ©W3C Working
Draft 30 April 2000, W3C (MIT, European Research Consortium for Informatics and Mathematics,
Keio University). All rights reserved

6.1.2.3 OpenVXI

Several tools can be downloaded from the Internet (at the time of writing) and used to develop
SDSs based on VoiceXML or CCXML. OpenVXI[18] is a library of open code that interprets
VoiceXML documents following the specification 2.0, avoiding proprietary extensions of
any kind. By using this specification, developers can benefit from the advantages offered
by well-known Web technologies and tools for the design of voice-based applications. The
interpreter is just a component of a complete VoiceXML platform, that includes simulated
ASR, generation of prompts and TTS, processing of user input, and basic telephony functions.
Its main advantage is that it provides some of the functionality required to execute dialogue
applications, such as an XML interface to process the VoiceXML documents, JavaScript,
etc. For the functionality concerned with the input/output, such as ASR, telephony control
and speech synthesis, it provides incomplete interfaces that can be changed to be adapted
to the specific platform features. In other words, it uses different ASR and speech synthesis
modules that can be created by the developers themselves. It is the developer's responsibility
to integrate all of these components.

The OpenVXI library was designed thinking about portability, and is appropriate for
browsers, testers, and debuggers of a great diversity of architectures. It was first introduced
by SpeechWorks International (at the time of writing ScanSoft), to accelerate the adoption of
the VoiceXML standard. Currently, the OpenVXI community includes some of the world's
largest manufacturers, software vendors, service providers, etc. Vocalocity[19] completed the
acquisition of OpenVXI from ScanSoft in August 2004, and offers several support levels
for current and future users. In addition to a GPL licence (General Public License) for non-
commercial use, it offers a commercial support licence for users interested in commercialising
and redistributing software.

---

[18] http://www.speech.cs.cmu.edu/openvxi/index.html
[19] http://www.vocalocity.net/

6.1.2.4 publicVoiceXML

publicVoiceXML[20] is an open-code browser that implements most of the functionality described in the VoiceXML 2.0 specification. It creates the links between the VoiceXML documents, the telephone line and the operating system. The browser uses an open-code VoiceXML interpreter provided by ScanSoft. To implement the telephone line functionality, it uses CAPI 2.0[21] over Linux (Red Hat 7.0 or 8.0, Debian 3.0 Woody, or Suse 8.0) or Windows (XP or 2000 Service Pack 2). TTS can be carried out using a third party's engine such as MS SAPI compliant (Windows), IBM ViaVoice or Festival. More than 20 languages are available (with Festival). In general, it is possible to use any telephony card compliant with CAPI 2.0. The supported telecommunication protocols are ISDN BRI or PRI, CAPI 2.0 and VoIP (with VoIPCAPI or XCAPY from a third party – only for Windows).

At the time of writing, the available version is 3.0, which includes speech synthesis via TTS, playback of audio files (only .wav), recording of user input (using .wav or .mp3 files) and DTMF for VoiceXML menus as well as `<field>` elements. ASR is not supported but there are projects to set it up (possibly in version 4.0). At the SourceForge page the reader can find additional information about development plans for publicVoiceXML 4.0, publicVoiceXML version 3.0, technical documentation, and mailing list, FAQ, configuration, use cases, etc.

6.1.2.5 IBM WebSphere Voice Toolkit

The IBM WebSphere Voice Toolkit[22] provides a full development environment that includes tools for setting up and debugging VoiceXML and CCXML applications, as well as tools to create and test grammars, build pronunciation files, and create and maintain natural language understanding (NLU) models.

The toolkit includes a graphic tool (called Call Flow Builder) that allows building new applications easily, using flowchart diagrams with nodes, links and decision elements, which simplifies notably the setting up of applications, specially for system developers who are not familiar with VoiceXML or CCXML (Figure 6.16(a)). The applications can also be built writing the corresponding VoiceXML or CCXML code (Figure 6.16(b)), using an editor with features for setting preferences, formatting, validation, etc. For example, the developer can use the preferences to set the DTD (Document Type Definition) used by the editor. The applications can be debugged using breakpoints, execution step-by-step, etc., placing calls using a Call Simulator. The toolkit includes wizards for converting applications written in VoiceXML 1.0 to VoiceXML 2.0.

The grammar tools define grammars of several types, as for example, JSGF, BNF, XML-based grammars, etc. These tools also interchange the format of the grammars, generate possible sentences, and check whether a sentence (in text or audio format) is allowed by a determined grammar. It also has a wizard to convert grammars from JSGF to SRGS.

The audio recorder records .wav or .au voice files that can be used as application prompts. The toolkit includes an XML-based editor for the lexicon, which creates pronunciation files

---

[20] http://publicvoicexml.sourceforge.net/
[21] http://www.capi.org/pages/downloads.php
[22] http://www-306.ibm.com/software/pervasive/voice_toolkit/

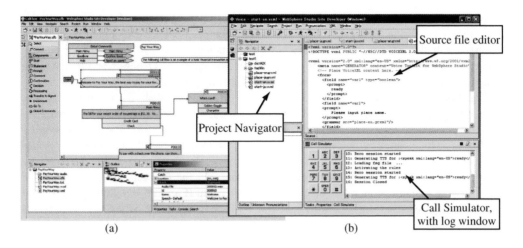

(a)   (b)

Figure 6.16 IBM WebSphere Voice Toolkit. Application development using: (a) Call Flow Builder. (b) Project Navigator, source code Editor and Call Simulator (for testing). (Reprinted by permission of 'Voice Toolkit for WebSphere' © 2004 by International Machines Corporation)

for ASR and TTS. It also creates pronunciation files in the format 'Pronunciation Lexicon Markup Requirements for the W3C Speech Interface Framework' defined by the W3C. Furthermore, it includes a tool that migrates pronunciation files created for previous versions of the toolkit to the new format. The toolkit also provides a tool for grammar checking based on MRCP (Media Resource Control Protocol).

The tools for the NLU development environment allow developers classify their data to generate several types of statistical models. Other developers can work simultaneously with the same data set, as they classify the way the NLU engines must process the answers provided by the user interacting with the application. Once properly classified, this data can be used to train new statistical models. These tools also create a DB2 database for the NLU development environment, validate the NLU models, import XML-based data into the database and export data from the database in XML format. The tools also provide several views to query the database, utilities to classify large amounts of data, and an editor for manipulating visually the Named Entity models as well as the parse trees.

The reusable components are modules already created to obtain information from the user, for example, a confirmation, a telephone number, a credit card number, etc., thus saving time and effort. There are two types of reusable components: *subdialogues* and *templates*. The former use predefined blocks to obtain a single piece of data from the user (e.g. a postal code) while templates obtain a set of related data, for example, in the case of an address, the postal code, street name and building number. The reusable components can automatically generate the necessary messages to obtain the data from the user, for example, predefined messages if there is no input from the user. They can also provide input examples and suggestions to spell the input or discard it. A wizard can be used to help configuring the parameters of the reusable components (e.g. the prompt to obtain the postal code).

At the time of writing, the Toolkit installation requires Microsoft Windows 2000 (Service Pack 4 or higher) or Microsoft Windows XP (Service Pack 1 or higher) as well as an installed version of one of the following: WebSphere Studio Site Developer 5.1.2 or WebSphere Studio

Application Developer 5.1.2 or WebSphere Application Server-Express 5.1.2. For ASR and TTS, the tool relies on the IBM WebSphere Voice Server Software Development Kit (SDK).

The installation selects several features to be installed: VoiceXML Application Development and Debug, Voice Portlet Application Development and Debug, Natural Language Understanding (NLU) Development, and Voice Tools for MRCP Sever.

The toolkit supports US English, UK English, Australian English, Brazilian Portuguese, Cantonese, Danish, French, French Canadian, German, Italian, Japanese, Korean, Norwegian, Simplified Chinese, Spanish (Mexico), Spanish (Spain) and Swedish. The speech synthesis for each language can be either concatenative or formant-based (the former provides more naturalness but requires more system resources). The recogniser and synthesiser for each language are installed at the same environment, but the language is set at the Voice Tools setting window, concretely at the *Preferences* menu (Figure 6.17). Setting a different language requires restarting the workbench after the new language is set, which prevents simultaneously using more than one language.

6.1.3 Internet Portals

Several Internet portals have been developed to host the development and test of SDSs based on some of these standard languages, for example, Voxeo Community, BeVocal Café and VoiceGenie Developer Workshop, which are discussed briefly in this section. Applications developed in most of these portal are accessed using VoIP (Voice over IP), which is a technology that allows making telephone calls over the Internet using a computer program (soft phone) or a physical telephone (hard phone) that understands the IP protocol. The soft phone is installed into the computer and runs on it as any other program. A variety of these phones can be downloaded for free from the Internet (e.g. Xten,[23] Pingtel,[24] eStara,[25] etc.).

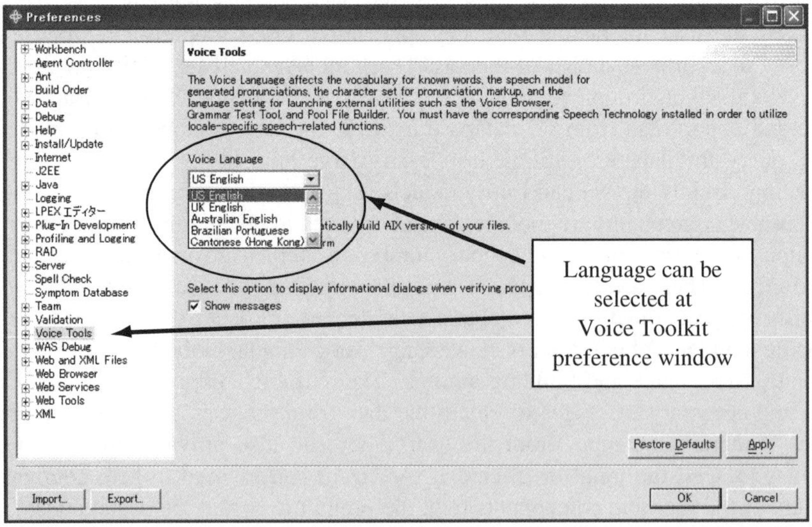

Figure 6.17 Language selection in the Voice Toolkit preference window (Reprinted by permission of 'Voice Toolkit for WebSphere' © 2004 by International Machines Corporation)

A hard phone is similar to a typical phone with the difference that it understands IP and some other protocols (SIP[26] and RTP[27]). VoIP phones connect the user to an IP-enabled data network, while typical phones connect to the PSTN (Public Switched Telephone Network). To use a soft phone, all that is needed is a direct connection to the Internet, a sound card, speakers and a microphone.

6.1.3.1 Voxeo Community

Voxeo Community[28] is a Web portal that provides several types of free resources to build and test SDSs based on VoiceXML, CCXML or CallXML.[29] The resources include:

- Tutorials for VoiceXML 2.0, CCXML 1.0 and CallXML 2.1 (see Figure 6.18).
- Pre-recorded audio-files for names of airplane companies, airports, letters, credit cards, months, week days, etc. (only for English). This material, recorded by The Great Voice

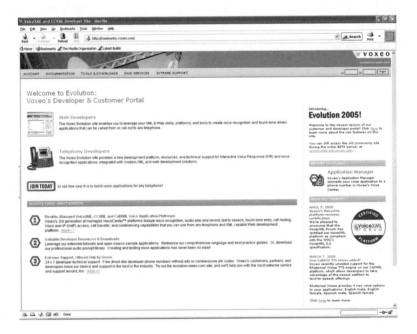

Figure 6.18 Voxeo Community developer site (reproduced by permission of Voxeo Corporation)

---

[23] http://www.xten.com

[24] http://www.pingel.com

[25] http://www.estara.com/softphone

[26] Session Initiation Protocol

[27] Real-Time Transport Protocol

[28] http://community.voxeo.com

[29] CallXML is another XML-based language similar to CCXML supported by the Voxeo Community, which allows user-system interaction using DTMF, robust error control, multi-conference, recording of audio messages, TTS and audio prompt generation.

Company, can be used for free by developers under an opensource LGPL (Lesser General Public Licence), and allows using high-quality voice in the applications.

- VoiceXML grammars for airplane companies, telephone area codes, city names, zip codes, numbers, week days, etc. ready to be used.
- Some sample applications that help create new applications as they show how to use a diversity of Web development languages. These applications are also released under the LGPL opensource licence.

All this material is free from the portal, even for non-registered users. Registration is required to develop and test applications, as well as to obtain help from an excellent 24x7 support team. Registered users (developers) have access to the application manager, application debugger, application monitoring and other tools.

Developers can interact with their applications via speech or DTMF, with telephone access offered in a number of ways. Voxeo offers access via a US 800 number, local numbers in several countries outside the US,[30] and SIP phone access over the Internet through providers like Free World Dialup (FWD). So, for example, after registering with Voxeo, a developer can use the portal to create an account on Free World Dialup (FWD), an Internet phone service that provides access via either a downloadable software phone, preconfigured for Windows,[31] or a preconfigured hardware telephone. Each new application is assigned either a telephone number (if accessing the application through a local US phone number provided by Voxeo) or a PIN code that can be used to access the application when using one of the other methods, such as the 800 number, the local numbers outside the US or FWD. The Voxeo Community offers 10 MB of free space in its own servers to store static applications, although developers can also use other Web servers (e.g. set in their own computers using Apache, Tomcat, etc.) to store their applications. Dynamic VoiceXML applications (based on ASP, JSP, Perl, ColdFusion, etc.) must be stored in the developers' own Web servers, which must be appropriately configured to support these languages (i.e. dynamic VoiceXML applications cannot be run in Voxeo Community Web servers).

The Voxeo Evolution portal uses the ScanSoft TTS engines, including both Rhetorical rVoice and Speechify, to offer male and female voices in English, Spanish and French. Voxeo also uses a variety of ASR technologies, including those from ScanSoft and Nuance, to offer speech recognition in English and Spanish.

A difference from other portals is that Voxeo Community supports the integration with CCXML, the W3C standard for robust call control. The portal also accepts/rejects incoming calls depending on the caller ID, as well as the transfer and redirection of calls, and the creation of multiconferences from 2 to 30 participants. Placing outbound calls from the portal requires that the developer has some privileges in his portal account. At the time of writing, these calls are charged to the developer credit card if their cost exceeds 100$ per month (at $0.18/minute). Outbound calls not exceeding that amount are free. Another feature of this portal is that the caller does not need to input a Developer ID and PIN to run the application, since every application (written in CCXML, CallXML or VoiceXML) is assigned a direct

---

[30] At the time of this printing, Voxeo offered local access numbers in the UK, Brazil, Italy, Spain, Israel, and Estonia, with announced plans for more numbers coming soon.

[31] The installation of the preconfigured Windows soft-phone carries out a calibration procedure and then creates an account in the Free World Dialup (FWD) network.

phone number. This feature saves time and avoids frustration, specially when the developer must call repeatedly to test the application. Finally, the portal also assigns failover URLs for the developed applications. This way, when an application is called, the Voxeo platform tries connecting the call to the first, second and third URLs specified (in this order) to ensure the call will be answered by the application.

6.1.3.2 BeVocal Café

BeVocal Café[32] also offers a diversity of tools to ease the development and test of Web-oriented SDSs based on VoiceXML 2.0 only, including the following tools:

- VoiceXML Checker: The VoiceXML Checker allows a developer to verify the correctness of VoiceXML files.
- File Management: Provides disk storage space and file management capabilities. VoiceXML resources can either be uploaded to Café or referenced via an external URL.
- Vocal Player: Allows the VoiceXML developer to replay all calls to their BeVocal Café account. This can be used for usability testing and grammar tuning when developing speech applications.
- Log Browser: The Log Browser allows the developer to view the log entries stored during each phone call and use it to debug their application using information about completed calls. The log browser provides the following classes of information: General Flow Trace, Event Trace, Performance Data, Variables Trace, Error Message, Custom Message, Call Control, Platform, Recognition Info, HTTP Header Info, Warning Message, and Timestamps.
- Vocal Debugger: This tool allows the developer to step through the source code and view the state of VoiceXML variables during the call. It helps them understand the flow of the VoiceXML application in a step-by-step manner.
- Trace Tool: This tool allows the viewing of log entries as they are generated during a call, and as one's application progresses.
- Vocal Scripter: Vocal Scripter provides a text channel to test the application flow of VoiceXML code. It decouples application flow testing from voice recognition testing and dramatically speeds up application testing, providing the ability to run VoiceXML applications using a text-based interaction. Interactive and batch modes are supported.
- Grammar Compiler: The Grammar Compiler allows the developer to submit a grammar source and have it compiled offline prior to executing a VoiceXML application. The application can then refer to the compiled grammar by using a special key provided after compilation. The Grammar Compiler supports Nuance GSL grammars. Compiling grammars before execution is useful with large grammars, where the compilation overhead during execution of a VoiceXML application could cause delays.
- Port Estimation Tool: The Port Estimation Tool estimates the number of telephony ports required by the application in order to provide a given quality of service.

The portal (Figure 6.19) also offers a variety of VoiceXML sample applications (e.g. an application that records what the user says and sends it to a server-side back-end, a four-function

---

[32] http://cafe.bevocal.com

calculator, an application to register and verify the user's voice print, etc.). It also offers several platform services samples based on Apache, Perl or Visual Basic (e.g. location information service, grammar compilation service, etc.). At the time of writing the interpreter supports US English and Spanish for ASR and TTS, and TTS for German. The interpreter does not support CCXML, but supports a subset of call control features via VoiceXML extensions.

To interact with the applications hosted in the portal, at the time of writing users can call 1-408-907-7328 or use the SIP URI sip:8773386225@voip.cafe.bevocal.com via a hardware or software-based SIP telephone. The recommended phone is the X-Lite Soft Phone from Xten Networks,[33] which is the same as the pre-configured software telephone installed from the Voxeo Community portal.

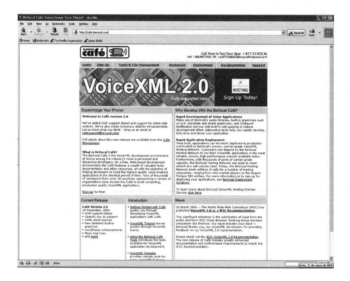

Figure 6.19   BeVocal Vocal Scripter developer site (reproduced by permission of BeVocal, Inc.)

6.1.3.3 VoiceGenie Developer Workshop

VoiceGenie Developer Workshop[34] is another portal that hosts the development and test of Web-based SDSs (Figure 6.20). It mainly differs from the Voxeo Community in that it only supports VoiceXML and the interaction with the developed applications must be carried out using an ordinary telephone. The portal supports two platforms for users: Developer platform and SpeechGenie platform. Each platform is assigned a different phone number (+1 416-736-9731 for the former and +1 416-736-6779 for the second).

The ASR and TTS supported languages depend on the platform used. In the case of the Developer platform, the ASR is based on Nuance 8.0 (for US English and German), while

---

[33] http://www.xten.com

[34] http://developer.voicegenie.com

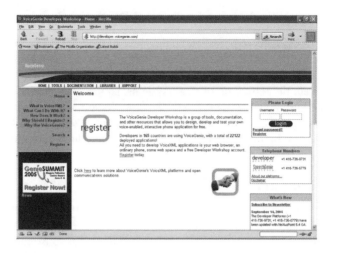

Figure 6.20 VoiceGenie Workshop developer site (reproduced by permission of VoiceGenie Technologies, Inc.)

TTS is based on SpeechWorks (now ScanSoft) Speechify 3.0 (for US English – female, and German – female).

In the case of the SpeechGenie platform, the ASR can be carried out using one of the following: SpeechWorks 3.0 Beta (for US English), Nuance 8.0 (for English), SpeechWorks 3.0 (for US English). SpeechWorks 1.1.3 (for US English, Australian English, Japanese). Each ASR engine requires a specific grammar format (if the application does not use grammars, any platform is suitable).

For TTS, one of the following languages can be used: SpeechWorks Speechify 3.0 (for US English male and female), SpeechWorks Speechify 2.1.6 (for US English male and female), Canadian French (male), German (female), American Spanish (female), Japanese (female), UK English (female), AT&T Natural Voices 1.4 (US English male and female), and American Spanish (female).

The portal provides tools for validation of VoiceXML files (either in the user's computer or in a URL), creation of different grammar types (GSL or XML in Nuance format, XML in SpeechWorks, now ScanSoft, format or ABNF), audio file conversion to Dialogic. Vox format, recording by phone of short audio prompts (maximum 30 sec) to be used in the applications, view of telephone calls' logs, conversion of VoiceXML files from one platform to another (e.g. files created for the Tellme Studio platform to the VoiceGenie platforms), etc.

The site also offers a 90-day trial version of an integrated development environment, called GenieIDE, that allows developing and testing applications locally and remotely (the user can also buy a registration key). The required OS is Microsoft Windows 2000/XP along with Service Pack 1. Enabling TTS for this environment requires installing the Microsoft Speech SDK 5.0.[35] Another feature is that the user can specify multiple ASR and TTS engines in the same script or application.

---

[35] http://www.microsoft.com/speech

6.2 Standards and Tools for Multimodal Dialogue Systems

The development of MMDSs requires integrating several modules that represent the diverse technologies involved (e.g. ASR, NLP, face localisation, gaze tracking, gesture recognition, fusion of multimodal data, etc.). Several tools have been made publicly available in the past few years to facilitate the setting up of these systems, so that non-specialist developers can implement simple multimodal systems relatively ease. Some tools develop isolated components (e.g. animated agents), whereas others develop working systems that allow using several input/output modalities. In this section we present a short introduction to some of these tools.

6.2.1 Web-Oriented Multimodal Dialogue

The W3C Multimodal Interaction Activity is making a lot of effort to develop standards for setting up systems than can be used to 'dialogue' with Web pages via multiple input/output modalities, such as speech, writing and 2D gestures. The goal is to extend the Web so that the users can dynamically select the most appropriate interaction modality. This way, e.g. he can enter a destination city either uttering the name (speech modality), using a keyboard to input the name in plain text (written modality), or encircling with a stylus a name shown on the screen (2D gesture modality). Analogously, the Web page can interact with the user visually and/or using speech.

An important difference between this Activity and the Voice Browser Activity is that, as commented in Section 6.1.2, the latter has a main specification (VoiceXML) and derivative specifications from it. However, in the Multimodal Interaction Activity there is not yet a specification for a central multimodal dialogue description language. Two current multimodal specifications are XHTML + Voice and SALT.

6.2.1.1 XHTML + Voice

XHTML + Voice (typically abbreviated X + V) is a markup language for multimodal Web pages that was first made public in 2001 as a result of the collaboration between IBM, Motorola and Opera Software, members of the W3C multimodal group. When interacting with an XHTML + Voice application, the user input can be done using speech, keyboard or a touch-sensitive screen and a stylus, whereas the system output is spoken and/or visual. The language supports the use of small devices (e.g. advanced cellular phones and PDAs) to access Internet contents. Moreover, the visual output compensates for the limitations of speech, given that the latter is not suitable to provide spatial information or information in the form of tables, which is more efficiently provided on the screen.

On the one hand, XHTML + Voice uses XHTML tags to determine the elements that will be seen by the user, the way he will see them and the action to carry out when he clicks on the screen or fills in a field by writing. On the other hand, the language uses VoiceXML tags to determine the sentences the user will be able to utter, their interpretation and the corresponding responses from the Web page. An application written in XHTML + Voice has four main parts: namespace declaration, visual part, voice part and processing part. The namespace is written in XHTML and includes declarations for VoiceXML and XML elements. The goal of the visual part (written in XHTML) is to define the diverse elements of the forms (e.g. fields, check boxes, etc.) that will appear in the device screen. The goal of the voice part (written in VoiceXML) is to generate prompts corresponding to the form fields and recognise the user sentences to fill in the fields. The processing part contains the event handlers, which include the instructions to execute when determined events occur.

Therefore, building an XHTML + Voice application requires making decisions for the visual part and the voice part (synthesised or pre-recorded messages, confirmation strategy, etc.). It also requires specifying the information that will be provided by each modality. The visual elements that are voice-enabled are text fields (that users can fill in either by writing or speaking), list boxes (in which users can select options using voice or the GUI), links to other Web pages and buttons in the current page. The application developer must take into account the synchronisation between both modalities, since the dialogue transitions will affect both modalities. For example, if the user clicks on a button to transit to another part of the application, the voice interface will need to stop the current processing and focus on the new point of interest. Analogously, when interacting using speech, the visual part must show on the screen the contents corresponding to the user utterance. The synchronisation is implemented using event handlers. XHTML + Voice includes XML standard event handlers for typical events, such as moving the mouse, placing the mouse over a menu option, etc.

For example, Figure 6.21 shows the multimodal pizza order form provided as one of the sample applications for the IBM Multimodal Tools (described briefly in Section 6.2.3). The user is presented with the visual dialogue box shown in the figure. The system speaks the initial prompt and the user can speak the required responses into a microphone by pressing the

Figure 6.21 Multimodal Pizza Order form (Reprinted by permission of 'Developing Multimodal Applications using XHTML + Voice' © 2003 by International Business Machines Corporation)

Scroll Lock Key as a 'Push & Talk' button. The user responses are inserted into the visual dialogue box and the system confirms the order when the 'Submit Pizza Order' button is pressed.

6.2.1.2 SALT

VoiceXML was designed to work only through the telephone line, therefore it does not use PDAs, Web browsers or other Internet-oriented devices. To overcome this limitation, the SALT Forum proposed in 2001 another XML-based language called SALT (Speech Application Language Tags) to access Internet contents using a variety of devices, such as Web browsers, PDAs and advanced telephones. The user interacts with SALT applications using speech, keyboard, keypad, mouse and/or stylus, whereas the application interacts with the user via speech, audio, plain text, motion video and/or graphics. For example, in an application that provides route information to travel between two points of a map, the system could provide several alternative routes connecting the two points, and ask the user using speech to choose one of them. The user could select the route by touching it on a touch-sensitive screen, or by saying 'I want the one on the right', for instance. The first working draft of the 0.9 specification was released in January 2002.

SALT also provides DTMF and call control for telephony browsers. It enables speech input for form filling scenarios using the 'Push & Talk' technique (studied in Section 2.1.1), or more sophisticated mixed initiative interaction, if needed. The speech recognition phase starts when the user presses with the pen on a textbox, which activates the grammar associated with the textbox. The recognition result is assigned to the textbox.

Several platforms have been developed to support application development using SALT, for example, Microsoft .NET Speech SDK 2.0 Beta and OpenSALT. The first is a set of tools based on the SALT specification that includes speech functionality in Web applications. It is integrated into the Microsoft Visual Studio.NET, including speech extensions for the Microsoft Internet Explorer browser. The download includes a great variety of sample applications and tutorials. OpenSALT is an open-code browser based on SALT 1.0 that, at the time of writing, is being developed at the Carnegie Mellon University (CMU) as an objective of the OpenSALT[36] project. This browser, based on the open-code Mozilla browser, will use the CMU open-code SPHINX speech recogniser and the Festival speech synthesis system (University of Edinburgh, UK).

As an alternative to using the tools included in the Microsoft Visual Studio, the developer can use any text editor to write the application files that contain the SALT tags, including all the assignments of the SALT attributes or properties, method calls and event handlers. Client computers can run the Web application using Microsoft Internet Explorer with the Microsoft Speech Add-in installed, which includes the DLLs that plug into the browser the capability to interpret the SALT tags.

SALT uses the same grammar specifications for ASR and for speech synthesis (SSML, Speech Synthesis Markup Language) as VoiceXML. Also, the core of SALT derives from other standards developed by the W3C, such as HTML, XML and XHTML, which facilitates the learning for developers familiar with these languages. In some aspects SALT is similar to XHTML + Voice; for example, it uses a small set of XML elements that are embedded into

---

[36] http://www.speech.cs.cmu.edu/openSALT

the host language used to write the Web page, such as XHTML, HTML + SIM or WML. Also, it uses an object-oriented event handling model that allows integrating several input modalities. SALT differs from VoiceXML in that the latter has a much bigger tag set, is used without being embedded in another language, and has its own flow control mechanism (the Form Interpretation Algorithm, FIA). On the contrary, in SALT the control flow is encoded in the host language, including the sequencing and the coordination of the events concerned with the prompt generation and speech recognition. Moreover, while in VoiceXML the acquired values during the interaction are automatically assigned to the fields, in SALT, the assignment must be explicitly coded, as happens in the case of XHTML + Voice.

6.2.2 Face and Body Animation

In the past few years, 3D animated agents have become very popular in industrial and research fields. The increase in computing power of personal computers allows state-of-the-art computer animation techniques to be used by end users. However, the lack of open-code tools to facilitate the development of facial animations limits the advances in this field, since each research group needs to set up its own development framework. Several efforts have been made to provide free and open-code tools to generate and animate 3D animated agents. In this section we introduce three of such tools: Xface toolkit, BEAT and Microsoft Agent.

6.2.2.1 Xface Toolkit

One of these efforts is the Xface toolkit (Balci 2004, 2005), an open-code platform[37] to design 3D animated agents based mainly on the MPEG-4 standard. Two of the main features of this toolkit are the ease of use and its extendibility. To generate the facial action parameters (FAPs) the toolkit relies on the *ampl2fap* tool (Lavagetto and Pockaj 1999) that processes APML (Affective Presentation Markup Language) scripts and generates the corresponding FAPs. APML easily defines the emotions and creates the associated animation parameters. For speech synthesis the tool uses another open-code tool: the Festival speech synthesis system.[38] The toolkit contains three main components. The core is the Xface library that allows developers include 3D facial animations in their software products. Moreover, researchers can implement their own working environments without needing to start from scratch. The XfaceEd editor (Figure 6.22a) is another component of the toolkit that easily generates MPEG-4 meshes from static 3D models. The third component of the toolkit is the XfacePlayer tool that observes the facial animation running (Figure 6.22b).

6.2.2.2 BEAT

The Behaviour Expression Simulation Toolkit (BEAT) allows the animator to enter typed text to be transmitted to the user, obtaining as output a synthesised message accompanied by non-verbal synchronised behaviour for an animated agent, which can be the input to

---

[37] http://xface.itc.it
[38] http://www.cstr.ed.ac.uk/projects/festival/

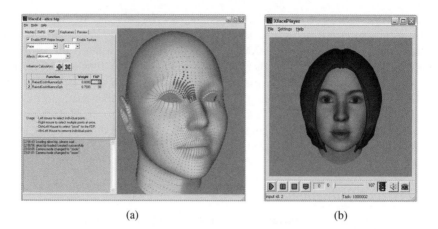

(a)   (b)

Figure 6.22   Xface Toolkit: (a) XfaceEd. (b) XfacePlayer (reproduced by permission of Balci 2005)

a variety of facial animation systems (Cassell et al. 2001). The analysis of the input text decides the movement of hands, arms, body and face, as well as the voice intonation, using a set of rules obtained from the analysis of human conversational behaviour. The toolkit was used to set up the animated agent of the Rea system, among others, and is extensible so that new rules can easily be added to decide the personality, movement and other features concerned with the facial animation. It was designed to be integrated with other systems that add personality profiles, movement features, or animation styles of a particular animator. The toolkit has been implemented to be modular and work in real time. It is based on a pipeline input/output structure that supports user-defined filters and knowledge databases. The processing is carried out by modules that work like XML-transducers, taking as input XML-tagged text and producing as output the same kind of tagged text. The input text to each module is transformed into a tree structure to be analysed, including information about non-verbal behaviour, and then transformed again to XML before it is sent to the following module. The knowledge databases are also coded in XML, which can easily be extended for other applications. The toolkit output specifies the agent behaviour, and contains tags that describe the type of animation to be performed and its duration.

6.2.2.3 Microsoft Agent

Microsoft Agent[39] (Trower 2000) is another tool that can be used to facilitate the setting up of interactive presentations using animated agents within the Microsoft Windows interface. It allows developers to integrate animated agents that behave as interactive assistants to introduce, guide, entertain or enhance Web pages and other applications, increasing the natural aspects concerned with human communication. The agent can generate speech using TTS or pre-recorded messages, can accept spoken commands, and can be combined easily with classic windows and menus to enhance the interaction with the Web page or application. It includes

---

[39] http://www.microsoft.com/msagent/default.asp

an ActiveX® control to make their services available to other programming languages such as Visual Basic® Scripting Edition (VBScript). Developers can use high-level functions to define typical movements of the agent, for instance, blink, look up or down, walk, etc.

Several authors have used this tool to include animated agents in their applications. For example, André et al. (1999) used the four predefined characters (Genie, Robby, Peedy and Merlin) in the Inhabited Mark Place system, whereas Prendinger et al. (2004) used it for setting up the Emphatic Companion, an animated interface that accompanies the user in a virtual job interview. Cavazza et al. (2001) also used it into a multimodal system to help users choose TV programs. The system interface uses the Microsoft Agent with a set of animated bitmaps acquired from a real human subject. The user-system interaction is synchronised with the non-verbal behaviour of the agent (i.e. her facial expressions) (Figure 6.23a) and the appearance of background images regarding the TV programs under discussion in a given moment of the dialogue (Figure 6.23b). In order to set up the animation of the agent, a human actress was filmed in advance. She was trained to adopt several facial expressions (e.g. happiness, sadness, surprise, etc.) and read some word sequences in order to obtain the necessary mouth disposition frames.

The agent's non-verbal behaviour was controlled using a small set of facial expressions regarding wonder, happiness and sadness, which represent an additional communication channel through which the user receives information without interrupting the dialogue flow. A set of local and global conditions determine the facial expressions. The local conditions take into account the user sentences considered individually; for example, a low confidence score or the impossibility of obtaining the semantic representation provokes a perplexed facial expression of the agent. On the contrary, if the user accepts a system suggestion, the agent makes a cheerful facial expression. The general conditions are concerned with the evolution of the dialogue considering the interaction history. These conditions make, for example, the agent look worried if the dialogue starts not to be productive.

6.2.3 System Development Tools

Several development tools can be found in the literature to facilitate the development of MMDSs. For example, Kamio et al. (1994) presented the Multimodal User Interface Design

(a)   (b)

Figure 6.23   (a) Non-verbal behaviour during the dialogue (b) The user interface with background images (reproduced by kind permission of prof. M. Cavazza, University of Teeside, UK)

Tool to support rapid prototyping of multimodal interfaces. In this tool, user interface objects are placed on a panel and links among the objects are used to describe plan-goal scenarios (what to do when specific events occur). The design tool then generates a script that drives the multimodal interface. Martin et al. (1995) presented a specification language for multimodal applications that describes the cooperation between modalities. For each task, the specification describes which modalities can be used to express determined parameters, and how these parameters are sent to an execution module. For each modality, a list of elemental 'events describe the information chunks that can be expressed in that modality, and how. Vo and Waibel (1997) presented a toolkit to develop MMDSs consisting of a set of grammatical tools that allow specifying multimodal applications using context-free grammars. The toolkit transforms automatically these grammars into the configuration files for the multimodal and integration modules of the system. The toolkit includes a graphic tool to design the grammars by drag-and-drop operations.

In the rest of this section we briefly describe two other tools than can be used to develop simple multimodal systems. The first (CSLU Toolkit) develops general-purpose systems whereas the second (IBM Multimodal Tools) has been designed to provide multimodal access to Web pages. The developed systems use multiple input/output modalities for the interaction, but do not carry out the fusion of input information chunks (studied in Section 2.2.1), e.g. they allow the user to enter a destination city using either speech, writing or pointing, but there is no fusion module that combines the information chunks provided through these modalities.

6.2.3.1 CSLU Toolkit

Since 1993 the Center for Spoken Language Understanding (Oregon Graduate Institute, USA) has been working on the development, acquisition and incorporation of speech technology in software environments that are easy to understand and use. The result of this effort is the CSLU Toolkit, a set of tools to learn and develop MMDSs and research on the technologies involved (McTear 1999, Cole 1999). The toolkit is free for download[40] for educational and research purposes and develops dialogue systems based on state-transition networks. It is also useful for data capture and analysis. The system input can be speech, DTMF or clicks on the screen, whereas the output can be speech, graphs or text, including an animated agent that can be selected by the user from several of them available (e.g. Gurmey, Lilly, Marge, Percy, etc.). The ASR can be based either on ANNs or HMMs, using a task-independent recogniser or specific recognisers for digits and alphanumeric strings. At the time of writing the available version for download is 2.0.0, which supports English (American and British), Spanish and Portuguese. The toolkit includes tools for administration (to set up user preferences, update through the Internet, etc.), alignment of speech files with the face movements of the animated agent (Baldi Sync and Multi Sync) and rapid application development (RAD). It also includes a command shell as well as documentation about RAD and the vocabulary tutor.

The main components of the RAD tool, shown in Figure 6.24(a), are the animated agent, the canvas, objects that can be dragged to the canvas, and a caption area where the messages

---

[40] http://cslu.ece.ogi.edu/toolkit/

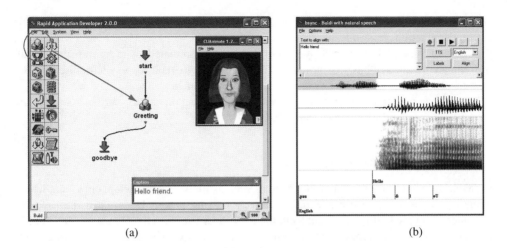

(a)   (b)

Figure 6.24   CSLU Toolkit: (a) RAD tool. (b) Baldi Sync tool (reproduced by permission of Oregon Health & Science University, Portland, USA)

generated using TTS (via de Festival speech synthesis system[41]) are shown in text format. To design a new application, the developer just needs to click and drag the necessary objects to the canvas and connect them appropriately with arrows.

The RAD tool includes a set of objects that allows the developer make use of several facilities for designing applications, as for example:

- Task-independent speech recognition, either isolated or continuous.
- Task-dependent recognition for digits and alphanumeric strings.
- Speech synthesis using either TTS or pre-recorded files.
- Communication media to generate sounds and place images on different areas of the screen.
- Sub-dialogues to avoid overloading the canvas with states and transitions (arrows) between states.
- Lists of associations between elements, e.g. queries-answers.
- Buttons on the screen to allow mouse-based user input.
- Tcl code to carry out sophisticated actions, e.g. searching for information in the Internet.
- Conditional and random flow transfer.
- User login by typing in a name.

The developer can define a variety of general preferences for the display (related to the animated agent, captioning, recognition results, canvas, etc.), data capture and playback, and miscellaneous (repair strategy, barge-in, etc.).

The CUAnimate tool (owned by the University of Colorado, USA) allows selecting the animated agent to be used in the application as well as some of its features, for example, face orientation, eyebrow and head movement, blink frequency, facial expressions, etc.

---

[41] http://www.cstr.ed.ac.uk/projects/festival/

(Ma et al. 2002). The tool uses full-bodied three-dimensional animated agents, and controls and renders them in real time.

The Baldi Sync tool refines the synchronisation of the animated agent lip movement with the sentence being synthesised through TTS (Figure 6.24(b)). It also synchronises the lip movements with natural voice pre-recorded by human beings.

Tutorials and sample exercises for developing simple applications can be found on the toolkit Web page and also in McTear (2004).

6.2.3.2 IBM Multimodal Tools

The wide-spread of mobile computing devices such as PDAs, mobile telephones, etc., which continuously decrease in size and increase in processing power, has created the need for new interaction modalities, different from the traditional ones based on keyboard and mouse. Current users demand access to information anytime and anywhere using these devices, thus making the traditional interfaces, e.g. based on HTML, inadequate to satisfy their needs and requirements in terms of performance and usability. In addition to using new devices, users demand that these devices allow them to interact multimodally so that they can choose the modalities that best fit their preferences and/or environmental conditions. The classical HCI (Human-Computer Interaction) paradigm in which the user must adapt himself to the requirements of the systems has changed: now the users demand that the systems adapt to their needs.

To facilitate the development, test and execution of multimodal applications that combine visual (based on XHTML) and spoken (based on VoiceXML) modalities, able to run on PDAs and other mobile devices, IBM has developed the Multimodal Tools,[42] which rely on the WebSphere Studio development framework.[43] This software includes two main components:

- *Multimodal Toolkit v4.3.2.* This toolkit extends the WebSphere Studio development environment providing an interface that creates applications to access Web contents multimodally, with significant savings of time and effort since developers do not need to know in detail the technologies involved in the setting up. The toolkit easily integrates the application's visual and acoustic parts, providing several tools, editors and views controlled by standard menus. The developer can use reusable components previously developed by IBM, which inserts automatically into the application the VoiceXML code necessary to obtain typical data using fields, such as postal address, credit card number, social security number, etc. When running the application, the data for each field can be provided in several ways.
- *WebSphere Everyplace Multimodal Browser v4.1.2.* A set of Web browsers developed in cooperation with Opera Software and ACCESS Systems that checks the performance of the XHTML + Voice applications developed using the Multimodal Toolkit v4.3.2. The browser is enhanced to allow the IBM ASR and TTS engines.

---

[42] http://www-3.ibm.com/software/pervasive/multimodal

[43] The installation of this software requires the previous installation of the WebSphere Studio Site Developer V5.1.2 or the WebSphere Studio Application Developer V5.1.2.

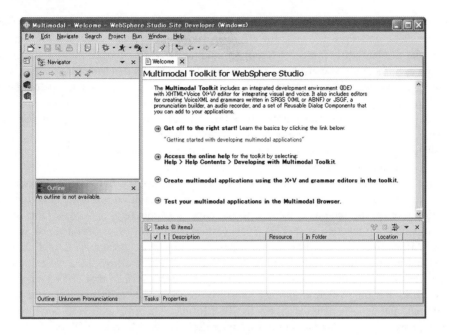

Figure 6.25 Multimodal perspective of the WebSphere studio (Reprinted by permission from 'Developing X + V Applications Using the Multimodal Tools' © 2003 by International Business Machines Corporation)

Figure 6.25 shows the Multimodal perspective of the WebSphere Studio development environment, which is comprised of four main panes: *navigator*, *outline*, *editor* and *tasks*. The *navigator* pane shows all the files associated with a development project. The *outline* pane shows, among other things, a window for the words with unknown pronunciation.

The *editor* pane is used to insert the multimodal application's XHTML + Voice code. As discussed in Section 6.2.1, the applications contain two parts, a visual one for pen/keyboard input (written in XHTML) and an acoustic one for speech input (written in VoiceXML).

6.2.4 Multimodal Annotation Tools

Multimodal corpora collected from the behaviour of human beings are very important for the research on MMDSs. Such corpora analyse the combination of the input modalities, as well as providing material to train the recognisers and obtain rules to model the behaviour of animated agents (Martin and Kipp 2002). In addition to multimodal corpora, algorithms are also necessary to process such corpora and quantify the observed multimodal patterns using significant measures that, among other questions, indicate how redundant is the users' behaviour, how frequently they use one input modality or another, or how frequently they change their interaction modalities. This information is very useful in analysing the multimodal behaviour and evaluating the need for multimodality in a particular scenario, as well as to evaluate an existing system by observing the relationship between multimodality and efficiency and/or acceptability.

Several tools have been developed to annotate and analyse multimodal corpora, and to compute these measures allowing for efficient data collection and analysis. For example, Müller and Strube (2001) presented a tool for annotating XML-encoded multimodal corpora in which each modality is considered as a stream of signals of a determined type (e.g. words, gestures, gazes, etc.) and as a set of user actions (e.g. button presses). Each type of signal is transmitted through a specific channel, thus allowing signals of different type be represented simultaneously in different channels. At a lower level, the tool uses a set of base level elements to represent word, gesture and key action signals, whereas at a higher level it uses two supra-base level annotation elements to put a structure over the base level elements in terms of turns and utterances.

Kipp (2001) presented Anvil (Annotation of Video and Spoken Language), a generic tool for the tagging of video that is also appropriate for the research on multimodal communication. The tool can be seen as a notepad to tag parallel events (e.g. a word, a sentence or a gesture) which are inserted at several levels (called *tracks*) together with a set of attributes using XML coding. The tool has several facilities to import/export data (PRAAT, XWaves, SPSS) and is extendable due to its plug-in architecture, which allows Java programmers to connect easily new modules. The tool also has a graphic interface that allows easy access to the track elements, whereas another interface observes and easily maintains the tagging carried out, and exports data massively in SPSS format. The tools is free to download[44] for research purposes and runs on Windows, Linux, Solaris and Macintosh OS X.

Martin et al. (2001) presented Tycoon, a tool to study and develop MMDSs based on a typology of the following types of multimodal cooperation: *equivalence*, *specialisation*, *complementarity* and *redundancy* (studied in Section 3.1.2). The tool includes the notion of *referenceable* objects, which are graphic objects that can be referenced by the user by speech or gesture. Using the tool, the tagging is carried out manually using a text editor and observing digital video tapes on a VCR. The evaluation measures are computed as a function of the percentage of the multimodal cooperation types found in the corpora.

As commented above, the analysis of multimodal corpora relies on the tags inserted by the analyser into several levels (also called *channels* or *tracks*) by applying a particular coding scheme. Appendix sets out several coding schemes for multimodal resources reported by Dybkjær and Bernsen (2002), which were collected in the European Natural Interactivity and Multimodality (NIMM) Working Group of the joint EU-HLT/US-NSF project International Standards for Language Engineering (ISLE). ISLE is the successor of EAGLES (European Advisory Group of Language Engineering Standards) I and II, and includes three working groups on lexicons, machine translation and evaluation, and NIMM, respectively. Many of the schemes mentioned in Appendix were created in the ISLE project, or by other researchers known by researchers participating in the project.

The ISLE MetaData Initiative (IMDI) created a corpus browser that permits a flexible and fast search for corpora and even sub-corpora. It is based on metadata descriptions which can be attached with human readable files to the corpora, making them available to everyone. These metadata include information about the project the resource belongs to, its creators, the media files and annotation available, its contents, the modalities and languages involved, etc. Additionally, the metadata can establish a corpus hierarchy which can be tailored to the

---

[44] http://www.dfki.de/~kipp/anvil/

user's particular point of view about corpora (Wittenburg et al. 2002b). The tools developed to support this infrastructure are a creator of IMDI metadata descriptions (IMDI BCEditor), a viewer of IMDI metadata descriptions (IMDI BCBrowser) and a search engine to query for the different resources (IMDI Search Tool).

6.3 Summary

In this chapter we initially discussed several tools for the development of SDSs and MLDSs, considering both the implementation of subsystems as well as the setting up of complete systems. Concretely, the chapter discussed a toolkit for implementing the ASR module (HTK), another toolkit for creating statistical language models for the ASR (CMU-Cambridge Statistical Language Modelling Toolkit), and four tools for setting up the speech synthesis (Festival, Flite, MBROLA and Galatea talk). Web-oriented standards and tools, including VoiceXML, CCXML open VXI, public VoiceXML and IBM were discussed. Then implementation tools, noting OpenVXI, publicVoiceXML and IBM WebSphere Voice Toolkit were outlined. In the next section some Internet portals for setting up SDSs (Voxeo Community, BeVocal Café and VoiceGenie Developer Workshop), which also provide information and tutorials about VoiceXML and CCXML, tools for developing and debugging applications, and different kinds of material to download (grammars, speech pre-recorded files, sample applications, etc.) were presented.

The second section of the chapter addressed the standards and tools for developing MMDSs. It initially focused on the Web-oriented multimodal dialogue, addressing briefly XHTML + Voice and SALT. Then it addressed tools to set up face and body animation (Xface toolkit, BEAT and Microsoft Agent). In the next section it focused on tools to set up complete MMDSs (CSLU Toolkit and IBM Multimodal Tools). To conclude, the chapter briefly discussed multimodal annotation tools (Anvil and Tycoon).

6.4 Further Reading

CCXML

- http://www.w3.org/TR/ccxml/
- http://docs.voxeo.com/ccxml/1.0/

CSLU Toolkit

- Chapter 7 and Appendix 2 of McTear (2004).
- http://cslu.ece.ogi.edu/toolkit/

Development of SDSs and MMDSs

- Part II of McTear (2004).

Grammar Formats

* Appendix 1 of McTear (2004).

IBM Multimodal Tools

* Appendix 4 of McTear (2004).
* http://www-306.ibm.com/software/pervasive/multimodal/

Standards and Resources

* Section 2.9 of Gibbon et al. (2000).
* Chapters 9–10 of McTear (2004).
* http://www.w3.org/TR/voicexml20/
* http://www.w3.org/TR/2003/WD-ccxml-20030612/
* http://www.saltforum.org/
* http://www.voicexml.org/specs/multimodal/x+v/12/

VoiceXML

* Chapters 9–10 of McTear (2004).
* Abbot (2002).
* http://www.w3.org/TR/2000/NOTE-voicexml-20000505/
* http://www.w3.org/TR/voicexml20

7

Assessment

7.1 Overview of Evaluation Techniques

Evaluation is a very important aspect in the technology of dialogue systems, since before setting a new system up in the real world, it must be evaluated to check its performance will be accepted by the users. The DARPA (Defence Advanced Research Projects Agency) in the USA was one of the precursory institutions for the establishment of periodic system evaluations. One of its more important projects, termed COMMUNICATOR, pays special attention to evaluation, considering user satisfaction as the main goal (Walker et al. 2001, 2002; Sanders et al. 2002). In Europe, some institutions concerned the evaluation of these systems are:

- COCOSDA (Coordinating Committee on Speech Databases and Speech I/O Systems Assessment), mainly concerned with the creation of multilingual databases (Mariani 1993).
- EAGLES (Expert Advisory Group on Language Engineering Standards), mainly devoted to the creation of a manual for developing linguistic resources and evaluating dialogue systems.
- ELRA (European Language Resources Association), mainly centred on the definition of a structure for the collection and distribution of linguistic resources.
- SQUALE (Speech Recognition Quality Assessment for Linguistic Engineering) devoted to the adaptation of the ARPA evaluation paradigm LVCSR (Large Vocabulary Continuous Speech Recognition) to a multilingual context (Young et al. 1997).

The evaluation of a dialogue system can be addressed following *subjective* or *objective* methods (Polifroni et al. 1998). The subjective (or qualitative) evaluation relies on the opinion of test users about the system performance obtained from interviews or questionnaires fulfilled by users after interacting with the tested system. An example of this technique is the so-called Trindi approach (Task Oriented Instructional Dialogue) developed by the Language Engineering sector of the European Union's program on Telematic Applications. According to this approach, the evaluation is carried out by answering affirmatively or negatively to a set of questions in a questionnaire. Thus, the more questions answered affirmatively, the more robust, flexible and natural is considered to be the system.

Spoken, Multilingual and Multimodal Dialogue Systems: Development and Assessment   Ramón López-Cózar Delgado
and Masahiro Araki   © 2005 John Wiley & Sons, Ltd

On the other hand, the objective (or quantitative) evaluation relies on mathematical computations that can be carried out following two different approaches, known as *crystal box* (or *subsystem*) and *black box* (or *end-to-end*) evaluation. In the former case, the evaluation is carried out considering the performance of each system's component (also called *subsystem*), while in the second the evaluation is carried out considering only system inputs and outputs.

In this section, we classify evaluation methods for SDSs, MLDSs and MMDSs, and discuss pros and cons of each method.

7.1.1 Classification of Evaluation Techniques

7.1.1.1 Subsystem-Level Evaluation

As can be observed in Figure 1.1, a dialogue system can be seen as a combination of input (e.g. ASR, NLU), processing (e.g. multimodal data fusion, dialogue manager, etc.) and generation modules (e.g. speech synthesis, graphic output, etc.). Since for some of these modules there are well-established evaluation methods, it is possible to evaluate the whole system by gathering the evaluation results for each component. Also, if the performance of the whole system depends largely on a specific module, the overall system performance can be estimated by evaluating the critical subsystem. For example, the overall performance of a car navigation system that receives as input spoken words depends largely on the word error rate (WER) of the ASR module. In the subsystem evaluation each module is assumed to be assessed independently from the others, which means that the system developer can find out the modules that must be enhanced. The drawback is in the difficulty of assessing the cooperation between subsystems. For example, the NLU module can absorb some ASR errors by employing a robust analysis technique, while the performance of the ASR module can be improved by predictive linguistic knowledge generated by the dialogue management module. The subsystem evaluation does not take into account this cooperation.

7.1.1.2 End-to-End Evaluation

An alternative method of evaluation is using the end-to-end evaluation, which uses predetermined 'outputs' that are compared with the outputs obtained by the whole system for the inputs. For example, in SDSs the input can be pre-recorded speech and the reference targets can be the sentences generated by the system. This method deals with the drawback of the subsystem evaluation in terms of not considering the collaborative behaviour between modules. However, it cannot measure the interactive ability of the system as it uses a predefined reference target for each dialogue turn. To evaluate interactive systems, pre-defined data cannot be used because the user's response determines what the system does next. The system's ability to resolve problems interactively or to recover from miscommunication cannot be evaluated by this method.

7.1.1.3 Dialogue Processing Evaluation

Another way of evaluating dialogue systems is to focus on the dialogue processing. The evaluation can be based on user-system experiments (including WOz evaluation) or analytical studies of the dialogue strategies. In the first case, the assessment is typically carried out using evaluation measures such as the task completion rate or time, and/or questionnaires filled out by the test users. Since this evaluation relies on a practical usage of the dialogue system, it can assess not only the collaboration between modules and the interactivity of the system, but also its usability, efficiency and friendliness. However, one drawback of this method is that it is only applicable when the system is fully implemented, i.e. it cannot evaluate the system at the intermediate development stages. Another drawback is that the evaluation takes into account human factors, which tend to be subjective and costly and require more time.

On the contrary, the evaluation based on an analytical model uses measures such as task completion rate, number of turns, turn-correction ratio, etc., considering other performance parameters of the system, such as WER, user misunderstanding rate, dialogue strategy, etc. The advantage is that this method is applicable at the system design stage, although sometimes it is difficult to model the behaviour of the dialogue processing using a simple formula.

7.1.1.4 System-to-System Automatic Evaluation

The suitable assessment method for a research system that often changes its behaviour is the so-called *system-to-system automatic evaluation* (Carletta 1992; Walker 1994a, 1994b; Hashida and Den 1999). Instead of human subjects, the evaluated dialogue system converses with another system. If the dialogue system is assumed to play an equal role to the human user in the dialogue, e.g. collaborative planning, another participant of the automatic evaluation can be the same system. Otherwise, the behaviour of human user is simulated by a computer program or pre-recorded data.

The input and output for each system are natural language text or artificial language (i.e. semantic representation). The dialogue is mediated by a *coordinator* program that opens the communication channel at the beginning of the dialogue, relays a message from one system to another in order to record a dialogue log, and closes the channel under certain conditions (e.g. explicit termination expression from one system, exceeding number of turns defined in advance, etc.). The log file is used to calculate task success rate, average number of turns, number of errors, etc. in order to evaluate the dialogue system.

In Japan, a competition of dialogue systems, termed DiaLeague, has taken place using this automatic evaluation method (Hashida et al. 1999). The task of the competition is to find out the shortest common route on slightly different maps (similar to railway route maps) which are incomplete. The maps consist of stations and connections, but some connections between stations might be inconsistent, some station names might be different, and some stations might be missing. The participating systems must find out the shortest routes from the source station to the destination station, exchanging information among them about the maps by using natural language in text format (Figure 7.1).

The aim of the competition is to measure the systems' ability to solve problems as well as their conciseness in the dialogue. Hashida et al. (1999) defined the ability of problem solving as the task completion rate, and the dialogue *conciseness* as the number of content words (thus, a smaller number of content words is preferable). This evaluation method can measure an overall performance and the interactive aspect of the systems. Although, in principle, it is only applicable to the dialogue system using written text, some researchers have extended to be used for SDSs.

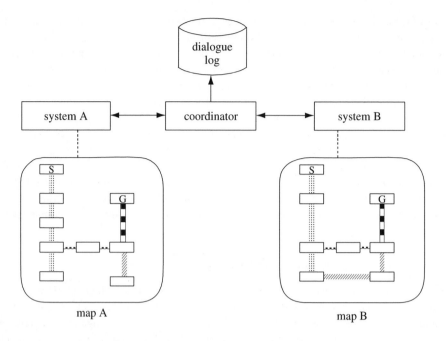

Figure 7.1 Coordinator-mediated dialogue between systems

7.2 Evaluation of Spoken and Multilingual Dialogue Systems

In this section we focus on evaluation methods for SDSs. Since multilingual systems are a particular case of SDSs, all the explanations in this section for SDSs are also applicable to multilingual systems. Exceptionally, a few researchers have addressed particular aspect of evaluation of multilingual systems, such as the design time. For example, D'Haro et al. (2004) presented strategies to reduce the design time of multimodal/multilingual systems considering three layers: (1) data modelling; (2) modality and language-independent dialogue flow modelling; and (3) modality and language-dependent interaction modelling. They showed how the clear separation of modalities and language-dependent information helped reduce the development time for multilingual/multimodal dialogue systems.

In accordance with the classification of evaluation methods presented in Section 7.1, we explain the Subsystem-Level, End-to-End, Dialogue Processing and System-to-System Automatic evaluation applied to SDSs.

7.2.1 Subsystem-Level Evaluation

In SDSs, the evaluation of recognition and understanding modules can be carried out using well-established methods that have been used for years of research on ASR and NLU. These methods are based on the concept of *input test set* with *reference target*. However, there are no well-established methods to evaluate the dialogue management and generation side modules. The difficulty in evaluating the dialogue management module is in the greater range of acceptable behaviour for this module, while the difficulty in evaluating the generation side modules is the inevitable involvement of human judges.

7.2.1.1 Evaluation of ASR

Well-established measures are typically used to evaluate the performance of the ASR subsystem. One measure is the so-called *word error rate*, or its opposite, the *word accuracy* (WA) which is calculated as follows:

$$WA = \left(1 - \frac{W_S + W_I + W_D}{W_T}\right) \times 100(\%) \tag{7.1}$$

where W_T is the total number of words in the reference corpus, W_S is the number of substituted words in the recogniser output, W_I is the number of inserted words in the output, and W_D is the number of deleted words in the output. Another measure is the so-called *sentence accuracy*, which is defined as follows

$$SA = (S_R/S_T) \times 100(\%) \tag{7.2}$$

where S_R is the number of sentences correctly recognised (i.e. the recogniser output matched exactly the input sentence) and S_T is the total number of sentences analysed by the recogniser.

For example, in the Julius toolkit (Japanese open source, real-time, large-vocabulary speech recognition engine) (Lee et. al. 2001), the recognition output file is analysed by a forced alignment tool (Figure 7.2) and several evaluation measures are calculated by a score calculation tool (Table 7.1).

In a large vocabulary task, the SA tends to be low partly because of a high rate of deletion errors of short functional words (e.g. 'at', 'of', 'in' in English, 'no', 'wo' in Japanese). Some of these errors are expected to be recovered to some extent by the process of converting the speech recogniser output into the semantic case frame.

Another measure that can be used to give a better estimate of the performance of the ASR subsystem, in connection with the sentence understanding subsystem, is a variant of the WA called *Keyword accuracy* (KWA), which takes into consideration only the words that are really necessary to obtain the correct semantics of the case frames, called *keywords*. Thus, this measure is defined as follows:

$$KWA = \left(1 - \frac{KW_S + KW_I + KW_D}{KW_T}\right) \times 100(\%) \tag{7.3}$$

where KW_S, KW_I and KW_D are, respectively, the number of keywords substituted, inserted and deleted by the recogniser, and KW_T is the total number of keywords in the input sentences.

It should be noted that WA, SA and KWA largely depend on the difficulty of the recognition task. In other words, as the vocabulary size and sentence complexity increase, the WA inevitably decreases. The so-called *perplexity* is an important factor to compare different recognition tasks, which measures the average number of branches at each word in the recognition process (i.e. the average number of words that can follow another word).

id: MA-001
REF: 京都駅 の 近く の 旅館 を 知りたい
HYP: 京都駅 の 近く　 旅館 を 知りたい
EVAL: C　　 C　C　　 D　C　　 C　C
CMSCORE: 0.986 0.997 0.871 0.907 0.985 0.134

id: MA-002
REF: 銀閣寺 の 拝観料 を 教えて 下さい
HYP: 銀閣寺 の 拝観料　　 教えて 下さい
EVAL: C　　 C　C　　　 D　C　　 C
CMSCORE: 0.870 0.997 0.529 0.994 0.987

id: MA-003
REF: 銀閣寺 の 行き方　　　 を 知りたい
HYP: 銀閣寺 の 行き方 の を 知りたい
EVAL: C　　 C　C　　 I　C　C
CMSCORE: 0.970 0.987 0.155 0.816

id: MA-005
REF: 銀閣寺 の　　 行き方 を 教えて 下さい
HYP: 銀閣寺 以下 京都駅　　 教えて 下さい
EVAL: C　　 S　　 S　　 D　C　　 C
CMSCORE: 0.993 0.914 0.524 0.434 0.972

REF: reference answer, HYP: hypothesis
C: correct, S: substitute, I: insert, D: delete
CMSCORE: confidence measure

Figure 7.2   Alignment of reference target and recognition hypothesis

Table 7.1   Example of summary of ASR results
SYSTEM SUMMARY PERCENTAGES BY SPEAKER

SPKR	Snt	Corr	Acc	Sub	Del	Ins	Err	S. Err
MA	10	74.24	74.24	4.55	21.21	0.00	25.76	100.00
Sum/Avg	10	74.24	74.24	4.55	21.21	0.00	25.76	100.00

Another evaluation aspect of the ASR module is the conformance with grammar description standards. This aspect becomes important in considering ease of development and re-usability of the module. The *de facto* standard of the speech recognition grammar is SRGS (Speech Recognition Grammar Specification).[1] This specification mainly indicates the format of the grammar definition, but also has some advanced characteristics such as weights for alternatives rules, garbage rules that may match any speech until the next rule match, the next token is found or the end of input sentence is reached. The standard representation of

---

[1] http://www.w3.org/TR/speech-grammar/

statistical language model[2] also is available, but many commercial or open source products that support statistical language model use each proprietary N-gram format at the time of writing.

7.2.1.2 Evaluation of NLU

The *sentence understanding rate* (SU) is a measure typically used to assess the performance of the NLU subsystem. It can be defined as follows, where S_U is the number of sentences correctly understood, and S_T is the total number analysed by the module.

$$SU = (S_U/S_T) \times 100(\%) \tag{7.4}$$

The sentence is regarded as understood if the output semantic representation coincides with the reference answer. The robustness of NLU module can be estimated by the difference between SU and SA.

Another factor to take into account is the subsystem ability for extracting the semantic information from partial parsing. For example, Funakoshi et al. (2003) applied their robust parser to the transcription of collected utterances in the simulated robot world. They classified the result of the syntactic analysis either as *correct* (when the obtained dependency-tree matched the speaker's intention, and the semantic slots assigned by the parser were also correct), *partially correct* (when the obtained tree was a subtree of the correct tree) or *wrong* (when the structure of the tree or the assigned semantic slots, or both, were inconsistent with the speaker's intention).

From the aspect of conformance of standardization, the output of this module should follow the semantic representation specified as EMMA (Extensible MultiModal Annotation markup language)[3] which permits a set of mutually exclusive interpretations and attaching confidence score of each interpretation. The output format is expected as an input for the dialogue manager in the Multimodal Interaction Framework.[4]

7.2.1.3 Evaluation of NLG

In general, generated text is evaluated by its comprehensiveness and naturalness. The generated sentences not only must be correct from a syntactic point of view, but also be correct within a given context, i.e. they must include elements such as ellipsis (to avoid unnecessary words) and use of pronouns instead of nouns, when possible.

Meng et al. (2003) evaluated their NLG module of SDS based on Grice's Maxims. Subjects of the evaluation are asked to fill out a questionnaire that consists of five questions: (1) Maxim of Quality; (2) Maxim of Quantity; (3) Maxim of Relevance; (4) Maxim of Manner; and (5) Overall User Satisfaction. Each question is represented as task specific question such as "Do you think that the answers of the virtual waiter are accurate and true?' The subjects are asked to answer these questions on a five-point Likert scale: very poor/poor/average/good/very good.

---

[2] http://www.w3.org/TR/ngram-spec
[3] http://www.w3.org/TR/emma/
[4] http://www.w3.org/TR/mmi-framework/

The human judge is indispensable in the evaluation of the NLG module. However, some questions are raised by this evaluation methodology. Does this evaluation suggest what is wrong in the NLG module? Is the generation process applicable to another task or domain?

7.2.1.4 Evaluation of Speech Synthesis

The goal of evaluating the performance of the speech synthesis module is to examine the quality of the synthesised voice. Since this evaluation is typically carried out by the human judge, one factor is the distribution of subjects who take part in the experiments, which must be balanced in terms of age, gender and experience in hearing speech synthesisers. Another factor is the synthesis unit (e.g. phonemes, diphonemes, triphonemes, syllables, words or sentences). The evaluation measures employed are generally concerned with clarity of the synthesised voice, comprehensiveness of content words, or naturalness of the synthesised speech.

The recent progress in standardisation of speech synthesis markup language has derived another possible criterion for evaluation: the conformance with a specification, such as SSML (Speech Synthesis Markup Language).[5] The specification requires prosody control, speaking rate control, emphasis of words or phrases, and appropriate text generation from some kind of expression (e.g. date, time, currency, abbreviation, etc.).

7.2.2 End-to-End Evaluation

Another evaluation method is the end-to-end evaluation for the evaluation of a unit (i.e. inter-related modules) in a SDS. This evaluation is carried out by comparing the input and output of this unit. The largest unit is the SDS itself which is evaluated by input and output speech. It is possible to set some intermediate unit in the SDS which is evaluated independently. For example, the *concept accuracy* (CA) (Boros et al. 1996) is a measure to evaluate the ASR and NLU modules as a whole, i.e. as a unit. The CA is defined very similarly as the WA, as can be seen in the following expression:

$$CA = \left(1 - \frac{C_S + C_I + C_D}{C_T}\right) \times 100(\%) \qquad (7.5)$$

where C_T is the total number of concepts (i.e. semantic units) in the reference target, C_S is the number of substituted concepts in the obtained output (taking into account the target), C_I is the number of inserted concepts in the obtained output, and C_D is the number of deleted concepts in the obtained output.

In general, the definition of semantic unit is different in task and domain of SDS. For example, Boros et al. (1996) defined semantic unit as an *attribute-value* pair that reflects the functionality of the system. As their system provides train timetable information, the semantic units are *sourcecity*, *goalcity*, *date*, etc. In the ATIS domain, SDSs can be evaluated by language input/database output pairs (Hirschman 1994). The interest for this evaluation is

---

[5] http://www.w3.org/TR/speech-synthesis/

the whole understanding process, in terms of getting the right answer for a specific task. The reference target is expressed as a set of minimal and maximal database tuples. The correct answer must include at least the information in the minimal target and no more information than in the maximal target.

7.2.3 Dialogue Processing Evaluation

As described in Section 7.1.1.3, there are two types of dialogue processing evaluation: evaluation by means of dialogue experiments with users (subjective evaluation) and evaluation by an analysis of the system's behaviour (objective evaluation). The former can be carried out using the PARADISE (Walker et al. 1997), and the second by an analytical approach. This section describes both methods. The latter approach is taken in a limited situation of the dialogue process.

7.2.3.1 PARADISE Framework

The PARADISE (PARAdigm for DIalogue System Evaluation) framework for evaluating SDSs uses user satisfaction ratings as an indicator of usability, and calculates the contribution of two potential factors to the user satisfaction (task success and dialog costs, as observed in Figure 7.3), using methods taken from decision-theory.

The task success is measured by using an *attribute-value matrix* (AVM) which is a kind of confusion matrix made from a dialogue corpus of the target system and human subjects. The AVM consists of a sub-matrix for each task slot whose entries are possible values for the slot. The value of the matrix is filled by the dialogue corpus with intended scenario

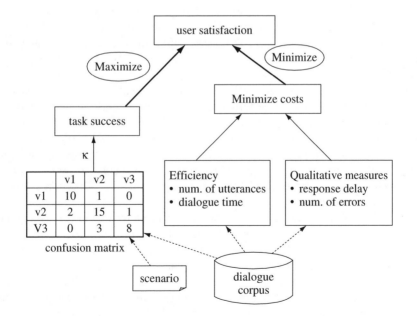

Figure 7.3 PARADISE framework

information that determines which is correct/incorrect. The task success is represented by the *kappa* coefficient (Carletta 1996) which is defined by the following expression:

$$\kappa = \frac{P(A) - P(E)}{1 - P(E)} \qquad (7.6)$$

where $P(A)$ is the proportion of times the AVMs for the dialogue data agrees with the scenario keys (i.e. correct value) and $P(E)$ is the proportion of times that the AVMs for the dialogue data agrees with the scenario keys by chance. Dialogue costs, which consist of efficiency measures and qualitative measures, are also acquired from the dialogue corpus. Thus, given a set of cost measures c_i, the system's performance is calculated as follows:

$$Performance = (\alpha * N(\kappa)) - \sum_{i=1}^{n} w_i * N(c_i) \qquad (7.7)$$

where α and w_i are weights that are determined by a multiple linear regression, and N is a normalisation function. Using this framework, the system performance can be evaluated at any level of the dialogue (i.e. general dialogue or specific sub-dialogue), using any dialogue strategies, for any tasks and domains.

7.2.3.2 Analytical Approach

The error recovery strategy in SDSs determines the overall system performance since ASR errors are unavoidable. The experimental evaluation, e.g. using PARADISE, can determine which strategy is the most suitable for a given situation. However, the amount of needed dialogue data is the product of the number of strategies. Thus, if several strategies consisting of a set of parameters are available for the system, the amount of dialogue data for the evaluation becomes unaffordable. Therefore, before implanting the dialogue strategies and carrying out dialogue experiments, an analytical modelling of them may give a good estimation of the dialogue cost.

Following this direction, Niimi et al. (1997) examined the relation between the efficiency of dialogue control strategies and the performance of a speech recogniser. Under the assumption that the system's goal is to collect and confirm several items (i.e. values of attributes) from the user and given the performance of the recogniser, they derived two quantities P_{ac} and N, which can describe the performance of the dialogue system using a proposed interaction strategy. P_{ac} is the probability that all information items included in user's utterance are conveyed to the system correctly, and N is the average number of turns taken between the user and the system until all the items are accepted.

In their method, the performance of the recognizer is given by a set of parameters shown in Figure 7.4. α, β, γ are the ratio of the user utterances (delivering information item) accepted, confirmed, and rejected respectively. q represents the ratio of correct recognition results under the condition of confirmed utterance (i.e. the confidence on the user utterance is below a confidence threshold and above a rejection threshold), δ is the probability that a response item is accepted and s is the probability that the accepted response has been recognised correctly.

Given these parameters, they show the mathematical derivation of the P_{ac} and N in different confirmation strategies.

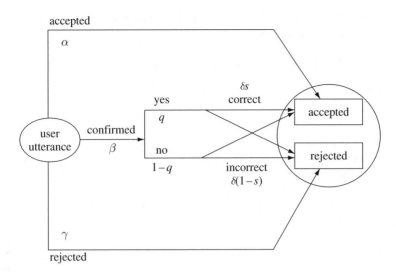

Figure 7.4 A set of parameters representing ASR performance (reproduced by permission of Niimi et al. 1997)

7.2.4 System-to-System Automatic Evaluation

The system-to-system automatic dialogue evaluation does not require a large dialogue corpus, and also checks the effects on dialogue efficiency and task completion of changing the dialogue strategies.

7.2.4.1 System-to-System Dialogue with Limited Resources

In evaluating dialogue strategies, the cognitive behaviour of the users is an important factor. From the collaborative point of view, the system should provide as much information as possible to contribute to the task achievement. For example, in a database search type task, if no data is obtained from a database query, the suggestions of how to relax a given conditions, frequently asked questions, recommendations of the record that do not exactly match the user query are helpful. However, considering the limited resources of users (e.g. capacity of working memory), short and concise system responses are favourable. Such a trade-off situation is suitable for the system-to-system automatic dialogue evaluation explained in Section 7.1.

Walker (1994a) examined the role of IRU (Informationally Redundant Utterance) of the resource-limited dialogue participant. According to the observation of the naturally occurring problem-solving dialogue corpus, around 12% of the utterances were IRU. She assumed that IRU helps the previously mentioned entity remember which may force a heavy cognitive load on the hearer if the IRU does not appear. In order to test the performance of several IRU usage strategies, she developed a test-bed environment, called Design World, where two dialogue agents (i.e. dialogue systems) collaboratively make a furniture layout in two rooms of a house. In Design World, the task requirement, agents' resources (i.e. size of working memory) and communicative strategies can be varied.

The performance of the dialogue agent in system-to-system dialogue with limited resources is defined by the following formula:

$$
\begin{aligned}
\text{Performance} = \ & \text{task defined raw score} \\
& - (\text{COMMCOST} \times \text{total messages}) \\
& - (\text{INFCOST} \times \text{total inferences}) \\
& - (\text{RETCOST} \times \text{total retrievals})
\end{aligned}
$$

where COMMCOST is the cost of sending a message from one agent to another in order to propose a layout plan, accept a plan, reject a plan, etc., INFCOST is the cost of means-end inference in making a proposal and RETCOST is the cost of retrieval the mentioned entity from the system's memory. The resource limitation of system's memory is simulated by a 3-dimensional grid space with the memory pointer and the search radius. For example, if the memory pointer is located at the grid $(0,0,0)$, the search radius of distance 1 is the set of points: $(0,0,1)$, $(0, 0, -1)$, $(0,1,0)$, $(0, -1, 0)$, $(1,0,0)$, $(-1, 0, 0)$. The *task defined raw score* is defined by the sum of the number of all the valid task steps in the agreed plan. If the dialogue agents pursue a perfect plan which gets a high task defined raw score simultaneously, they must consume a large amount of cognitive costs.

She showed that strategies that seemed to be inefficient under assumptions of unlimited-resource agent are effective with resource-limited agents under certain conditions. Furthermore, she indicates that different tasks make different cognitive demands, and place different requirements on the agents' collaborative behaviour. Complex task settings and fault-intolerant tasks can benefit from communicative strategies that include relatively frequent IRU.

7.2.4.2 System-to-System Dialogue with Linguistic Noise

Araki and Doshita (1997) extended the concept of system-to-system automatic dialogue for the evaluation of SDSs. As can be observed in Figure 7.5, the evaluation uses random linguistic noise that is put into the communication channel by a dialogue coordinator in order to simulate speech recognition errors. The point of the evaluation is to judge the subsystems' ability to repair or manage the misrecognised sentences by using a robust linguistic processor or by using the appropriate dialogue management strategy.

The evaluation environment consists of a coordinator agent and two dialogue agents. At the start of the dialogue, the coordinator sends a start signal to one of the dialogue agents, and the agent that receives the start signal (system simulator) opens the dialogue. The system simulator generates natural language output in text format which is sent to the coordinator. The coordinator receives the text, puts linguistic noise into it at a given rate, and passes the result to the other dialogue agent (the user simulator), which has the next turn. The dialogue ends when one of the dialogue agents cuts the connection or when the number of turns exceeds a given upper bound.

The system performance is measured by the task achievement rate (ability to solve problems) and the average number of turns needed for the task completion (conciseness of dialogue) under a given WER. The result of the dialogue is examined using the logged data. If both agents achieve the same and correct conclusion, we regard the task problem as solved. The task achievement rate is calculated from the number of dialogue trials that reach

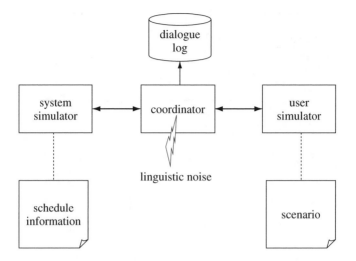

Figure 7.5   System-to-system dialogue with linguistic noise

the same and correct conclusion divided by the total number of dialogue trials. In addition, the evaluation assumes that the conciseness of a dialogue can be measured by the average number of turns.

Using this environment, several dialogue strategies can be examined to answer the following questions: What type of feedback and error recovery techniques are suitable for a given WER? What level of initiative is suitable for a given situation? Such dialogue strategies are represented by a set of parameters (e.g. level of initiative, type of feedback, frequency of confirmation, etc.) in the system simulator. By changing these parameters and running the experiments again, the most suitable settings for these parameters can be found.

7.2.4.3 System-to-System Dialogue with a User Simulator

The above system-to-system dialogue simulations can be used to evaluate dialogue strategies without considering real speech recognition errors. In order to avoid this drawback, López-Cózar et al. (2003) proposed a method that re-uses a corpus of sentences uttered by real users in order to evaluate a SDS implementing different dialogue strategies. These sentences, in the form of voice signal files, are analysed by the system's speech recogniser. The method is based on the system-to-system approach; concretely, it relies on the automatic generation of conversations between the SDS to be evaluated, and an additional 'dialogue' system called *user simulator* that represents the user interacting with the SDS. The evaluation environment is shown in Figure 7.6.

The method was applied to evaluate a SDS designed for the fast food domain under two different recognition front-ends and using two different dialogue strategies to handle user confirmations. The experiments showed that the prompt-dependent recognition front-end achieved better results only if the users limited their utterances to those related to the current system prompt. The prompt-independent front-end achieved inferior results, but

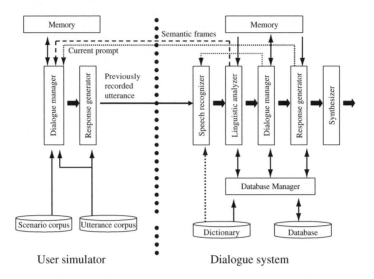

Figure 7.6 System-to-system dialogue using a user simulator

enabled users to utter any permitted utterance at any time, irrespective of the system prompt. In consequence, this front-end allowed a more natural and comfortable interaction. The experiments also showed that the re-prompting confirmation strategy enhanced the system performance for both recognition front-ends.

7.3 Evaluation of Multimodal Dialogue Systems

The evaluation of MMDSs is a open field given the current lack of standards and methods clearly accepted by the research community. There is also a lack of available multimodal databases that can be used as a reference in the evaluation, given that those used to evaluate the diverse modules (e.g. speech recogniser, handwriting recogniser, face locator, etc.) are not suitable to evaluate one of the main aspects of multimodal systems: the combination of the diverse interaction modalities. For example, users may enter queries using speech and provide additional or more specific information by pointing and/or by making gestures, thus making more complex the interaction and thus the evaluation. Additionally, the evaluation of qualitative aspects, mainly concerned with the system performance as a whole, plays a very significant role in multimodal systems. This is in clear contrast to the evaluation of isolated components in which the evaluation is typically made quantitatively, mostly using accuracy in diverse forms (e.g. word accuracy in the case of ASR).

Also, measuring the contribution to the system performance of the different recognition modules is not as straightforward as in the case of SDSs, since these modules deal with very different tasks in terms of complexity. For example, typically the speech recogniser carries out a very difficult task whereas the gesture recogniser deals with a very small set of possible gesture to recognise. Consequently, the gesture recogniser will be more reliable than the speech recogniser, but the former must have a higher weight when evaluating the system performance, given that its task is much more complex. In addition to using different weights for the different recognisers, the evaluation of multimodal systems must face a problem of synchrony since it is necessary to decide which multimodal inputs must be considered synchronous (i.e. referring to the same user intention) and which ones are equivalent (i.e. describing the same user intention) although not synchronous in time.

The growing sophistication of MMDSs requires new evaluation measures since, among other considerations, these systems can be implemented in mobile devices, which present specific problems for the evaluation that still have not been completely resolved. For example, some issues are *how* and *when* to use the very small displays of these devices in combination with speech, for which application types a system should use information about the user localisation in the environment, in which situations using the displays may be considered (in)secure (e.g. while driving a car), etc. (Bühler et al. 2002). Also, the evaluation of emotional systems (Section 3.2.5.1) requires taking into account new aspects not considered previously, as for example the impact of the emotional models employed by these systems on the user. The specific characteristics of MMDSs provoke a lack of evaluation methods that can be considered *standard*. Thus, in this section we report an overview of current trends, existing experiences and results in the evaluation of such systems. Of course, the overview may be far from being complete due to space limitations. As in the case of SDSs, two types of evaluation can be considered for MMDSs: *system-level* and *component-level*.

7.3.1 System-Level Evaluation

Several evaluation approaches taken from the field of HCI are typically applied to the system-level evaluation of MMDSs depending on the type of multimodal data fusion used. In the case of signal-level fusion, the evaluation measures generally used are similar to those employed to evaluate the isolated components whose outputs are combined in the fusion process. For example, in the case of AVASR the evaluation measures are some form of recognition accuracy, since it is a typical evaluation measure for speech and lip-reading recognisers. In the case of semantic fusion, the evaluation measures generally used are task completion rate, task completion time, naturalness, user satisfaction, costs, etc. Among others, Cerrato and Ekeklint (2002) used user satisfaction, taking into account the way users finished the interaction with the system; concretely, they considered that prosodic aspects (e.g. tone envelope and intensity) can be indicators of this measure. Their results show that there is a user trend to finish the interaction in a 'conventional way' if the interaction is carried without problems and with mutual understanding. The analysis of the acoustic measures obtained from the farewells and greetings show the prosodic aspects can effectively be used as indicators of user satisfaction. Other authors have used as evaluation measures *work load* and *user preference* in selecting the interaction modalities. For example, Woltjer et al. (2003) evaluated an e-mail in-car multimodal system comparing several interaction and presentation modalities, finding that the subjective workload was higher when the combination of manual/visual modalities was used. When the test users were free to select the modalities, the workload was considered lower whereas the performance and preference were considered the highest for the conditions tested.

Three approaches to system-level evaluation can be considered: experimental, predictive and expert-based.

7.3.1.1 Experimental Approach

The experimental approach focuses on real data obtained from real users carrying out real tasks. The evaluation usually relies on questionnaires filled out by users after they interact with simulated, prototype or final systems. As in the case of SDSs, the evaluation of multimodal systems generally relies on scenarios that define a set of goals the users must try to achieve when they interact with the dialogue system (Koda and Maes 1996; Sturm et al. 2002; Heylen et al. 2002; Bickmore and Cassell 2004). The questionnaires are used to discover their opinion after the interaction in terms of several subjective measures, for example, ease of use, satisfaction, involvement in the task, efficiency, comfort during the interaction, etc. In MMDSs, some additional measures are personality/character of the animated agent, naturalness of the animated agent, number of errors and speed of the different interaction modalities, etc. These subjective measures are generally mapped to affirmations with which the test users must agree or disagree using a Likert scale (Table 7.2 shows some examples). For instance, on a five-point scale: 'I disagree' $= -2$, 'I agree' $= -1$, 'Indifferent' $= 0$, 'I agree' $= 1$, 'I strongly agree' $= 2$. A variety of statistic methods are used to calculate the values for these measures considering the answers to the questionnaires, for example, ANOVA (Heylen et al. 2002; Krum et al. 2002; Bickmore and Cassell 2004), Tukey post hoc (Krum et al. 2002), Kruskal-Wallis (Heylen et al. 2002), Mann-Whitney (Fabri et al. 2002), etc.

For example, Sturm et al. (2002) used questionnaires to compare the performance of the MATIS multimodal system with two other versions of the same system: a graphic-only version accessible through the Internet and a speech-only version accessible through the telephone line. Test users carried out three different interaction scenarios in which goals

Table 7.2 Mapping subjective measures-statements for subjective evaluation

Subjective measure	Statements
Ease of use	It was easy to get the information.
	It was easy to carry out the task.
	It was clear what to ask or say.
Satisfaction	I liked using the system.
	The system took a long time to respond.
	The dialogue had a clear structure.
	I liked carrying out the task using the system.
Involvement	I looked at the animated agent about as often as I use to look at interlocutors in normal conversations.
	The animated agent kept the distance.
	It was clear when the animated agent finished speaking.
Efficiency	It took a reasonable time to carry out the tasks.
Personality/character of the animated agent	I trusted the agent.
	The agent was a friendly 'person'.
	The agent was bad-tempered.
Naturalness of the animated agent	The movements of eyes, lips, head and other parts of the body of the agent seemed natural.

were expressed graphically to avoid interfering with their own way of expressing. After the three versions of the system were tested, using a Likert scale, users ranked their opinions in questionnaires in terms of system effectiveness and efficiency, user satisfaction and preference for future use. The evaluation results showed the multimodal version resolved some problems of the spoken version, providing a more effective and efficient interaction. In general, users indicated they preferred the multimodal version due to its visual features, but those users who had a lot of experience interacting with GUI interfaces clearly preferred the graphic-only version, since they found it faster and least error-prone.

If the evaluation is performed using a prototype system, the experiments obtain a great amount of data varying from quantitative measures to informal observations, which can be very useful to create multimodal databases usable in further evaluations. For example, Johnston et al. (2002) used a prototype version of the MATCH (Multimodal Access To City Help) system that provided restaurant and subway information using speech and pen input. Using this prototype, the test users could say 'Show me cheap Italian restaurants in Chelsea', or enter the same command multimodally by circling an area on a map while saying 'Show me cheap Italian restaurants in this neighbourhood'. Table 7.3 shows the results obtained for five subjects (2 male, 3 female) not involved in the development of the prototype who performed a total of 338 interactions divided as follows: 171 (51%) were speech only, 93 (28%) were multimodal, 66 (19%) were pen only, and 8 (2%) were GUI actions.

As can be observed, even though the word accuracy was low, the task completion rate was high, which indicates that by integrating the multimodal aspects of the system the users were able to complete the tasks successfully.

As in the case of SDSs, it may require a lot of effort to create a fully functional prototype system that can be used to carry out user studies. A solution to this problem is to simulate the components of the system that have not been developed and evaluate the system using the simulated components. The WOz technique, discussed in Sections 3.2.1.1 and 3.2.1.2, has been widely used for this purpose.

7.3.1.2 Predictive Approach

The predictive approach judges the user behaviour considering performance variables that take into account suppositions or parameters of a model, without using a previously implemented system nor user interaction (Mellor and Baber 1997). Some predictive techniques are CCT (Kieras and Polson 1985), ICS (Barnard 1987), KRI (Löwgren and Nordqvist 1990) and cognitive walkthrough (Lewis et al. 1990). The main advantage of this approach is that it evaluates the user interface at an early stage of the development process, so that the design can be enhanced before the implementation takes place. The disadvantage is that the predictions are based on hypothetical theories and not on real data, thus these techniques

Table 7.3 MATCH performance evaluation

ASR word accuracy	59.6 %
ASR sentence accuracy	36.1 %
Task completion time	85 %
Average time/task	6.25 m

may be little precise if the case under study (e.g. multimodal interaction) is not supported by an underlying theory. Another drawback is that specifying a predictive model might take almost the same time as implementing a prototype system.

7.3.1.3 Expert-Based Approach

In the expert-based approach, an experienced professional uses a prototype system and evaluates its specification in a more or less structured way to determine whether it complies with predefined design criteria. The main drawback of this approach is that it is difficult to find the expert, using evaluators instead to find out a reasonable number of design problems (typically at least three evaluators are necessary to find out half of the problems). Among others, this approach has been used by Almeida et al. (2002) to carry out the evaluation of the MUST (Multimodal, multilingual information Services for small mobile Terminals) system, which uses speech and pointing for input and speech, text and graphics for output. The system was set up in a mobile device to provide tourist information using maps that not only showed the location of streets, but also displayed graphics of main buildings and monuments. The evaluation results, carried out by 12 HCI experts, showed that most experts started interacting using the two input modalities separately, and some of them even never tried to combine them. After some time using the system, five experts started to use speech and pen input simultaneously. Thus, the experiments showed that it was not intuitive nor obvious that the system was multimodal, and that both input modalities could be used simultaneously, suggesting that for novice users it would be necessary to give an introductory explanation of the service and the interface, e.g. using a video or a short animation.

7.3.1.4 The PROMISE Evaluation Framework

Given the specific features and the problems in evaluating MMDSs, Beringer et al. (2002a) proposed the PROMISE (PROcedure for Multimodal Interactive System Evaluation) evaluation framework, which considers methods used to evaluate SDSs but also new methods to take into account specific features of MMDSs, for example, the combination of speech and gestures in the input, the combination of speech and graphics in the output, etc. Using this framework, the evaluation is carried out subjectively by defining qualitative and quantitative measures (called *costs*) which have weights associated. Instead of using a lineal regression (used in the PARADISE framework explained in Section 7.2.3.1) the PROMISE framework uses a Pearson correlation that is calculated between pairs 'User satisfaction – cost'. To carry out the evaluation, test users interact with the system and fulfil a questionnaire which includes measures that can be rated objectively. For example, Tables 7.4 and 7.5 show the quality and quantity measures, respectively, used to evaluate the SmartKom system using this framework. Some of these measures are equivalent to the measures used in the PARADISE framework, whereas others are defined to deal specifically with multimodality and non-cooperative users.

In addition to quality measures such as task completion rate, response time, number of turns, etc., this framework uses a cost related to the score of the multimodal inputs and outputs (the 'semantics' cost), which measures the number of multiple input and the possible

Table 7.4 PROMISE quality measures (reproduced by permission of Beringer et al. 2002a)

System-cooperativity	Measure of accepting misleading input
Semantics	No. of multiple input possible misunderstandings of input/output semantical correctness of input/output
Helps	No. of offered help for the actual interaction situation
Recognition	Speech Facial expression Gestures
Transaction success	No. of completed sub-tasks
Diagnostic error messages	Percentage of error prompts
Dialogue complexity	Task complexity (needed information bits for one task) Input complexity (used information bits) Dialogue manager complexity (presentation of results)
Ways of interaction	Gestures/graphics vs. speech N-way communication (several modalities possible at the same time?)
Synchrony	Graphical and speech output
User/system turns	Mixed initiative Dialogue management Incremental compatibility

misunderstandings of input/output due to the multimodal interaction. The framework uses weights for the different recognisers, obtained via the Pearson correlation. To evaluate the system (in terms of performance) the framework makes a generalisation of the diverse costs c_i using the normalisation function:

$$N(c_i) = \frac{c_i - \overline{c_i}}{\sigma_{c_i}} \qquad (7.8)$$

where $\overline{c_i}$ and σ_{c_i} are the cost average and variance, respectively. Since costs c_i are not equally likely, the framework uses weights w_i that correlate 'user satisfaction – cost' via the Pearson correlation function. Considering n costs, the performance is defined as follows:

$$performance = \alpha\overline{\tau} - \sum_{i=1}^{n} \varpi_i N(c_i) \qquad (7.9)$$

where α is the Pearson correlation between the average task success[6] $\overline{\tau}$ and the corresponding user satisfaction values.

---

[6] The average task success $\overline{\tau}$ is calculated from the τ_j's where j is the index of tests: $\tau_j = +1$ if task success; $\tau_j = -1$ if task failure.

Table 7.5 PROMISE quantity measures (reproduced by permission of Beringer et al. 2002a)

Barge-in	No. of user/system overlap by means of backchannelling
	Negation of output
	Further information
Cancels	Planned system interrupts due to barge-in
Off-talk	No. of non-system directed user utterances
Elapsed time	Duration of input of the facial expression
	Duration of gestural input
	Duration of speech input
	Duration of ASR
	Duration of gesture recognition
	Mean system response time
	Mean user response time
	Task completion
	Duration of the dialogue
Rejections	Error frequency of input which require a repetition by the user
Timeout	Error rate of output
	Error rate of input
User/system turns	No. of turns
	No. of spoken words
	No. of produced gestures
	Percentage of appropriate/inappropriate system directive diagnostic utterances
	Percentage of explicit recovery
	Answers

7.3.2 Subsystem-Level Evaluation

For the component-level (or *subsystem-level*) evaluation, researchers have traditionally used measures in the evaluation of the diverse recognisers of the system (for speech, handwriting, gestures, etc.) as well as specific measures for evaluating animated agents. In this section we present a short overview of the measures used for evaluating face localisation and gaze tracking, gesture recognition, and handwriting recognition. The evaluation of the multimodal data fusion and the animated agents, although carried out at the system-level, are addressed with more detail in Sections 7.3.3 and 7.3.4, respectively.

7.3.2.1 Evaluation of Face Localisation and Gaze Tracking

Given that face localisation requires a previous detection step, an evaluation measure generally used is the face detection rate (ratio of correct detections to all human faces in a corpus). For example, Rowley et al. (1998) used this method to evaluate a face detection system based on an ANN, obtaining 90.5% detection rate of frontal view faces when processing a corpus that contained 130 grey-scale images. Similarly, Hsu et al. (2002) evaluated a system to detect faces in colour images considering the same evaluation measure, but also the average

CPU time required to process the images. Related to face detection, Yang et al. (1998) evaluated the estimation of the user head pose in terms of accuracy, considering the average rotation error (measured in degrees) and the average translation error (measured in mm.).

After the user face has been located, it can be tracked. Several evaluation measures have been used to track faces and other facial features. Tracking accuracy (percent deviation from true position) and tracking success (ratio of time when the feature is tracked and when it is lost) have been perhaps the most used. For example, Yang et al. (1998) evaluated a gaze tracking system in terms of accuracy (average error measured in degrees). Gaze tracking has also been evaluated in terms of the time it requires to carry out a specific task in comparison with that required by other interaction modalities. For example, Sibert (2000) compared the use of gaze tracking and mouse in an object-selection task, finding that the gaze selection was 338 ms faster.

7.3.2.2 Evaluation of Gesture Recognition

As noted in Section 2.1.6, many methods have been proposed for gesture recognition, mainly based on templates, ANNs, HMMs, Bayesian Networks, Principal Component Analysis and artificial vision techniques. Gestures may represent a single command, a sequence of commands, a single word, or a phrase, and may be static or dynamic. The evaluation of recognition systems is typically carried in terms of accuracy (i.e. proportion of correctly recognised gestures), although some researchers have also focused on recognition time and robustness (Hu et al. 2003, Shamaie and Sutherland 2003, Wachs et al. 2002).

A diversity of results can be found in the literature obtained for a diversity of applications, as for example interactive maps, interaction in office environments, interaction with robots, etc. For example, experiments conducted by Kettebekow and Sharma (2001, 2000) in the weather domain and the iMAP framework reported 78.1 and 79.6% recognition rates, respectively. In office environment applications, Montero and Sucar (2004) studied a set of gestures used to interact with other objects, for example, a mouse, a drawer, etc. Hand gestures were detected and tracked using a camera and adaptive colour histograms. The recognition was based on HMMs, considering several number of hidden states and combinations of features (e.g. velocity, Cartesian coordinates, etc.). The results showed a great variability on recognition rates depending on these parameters, from less than 50% to more than 95%.

In the field of robot interaction, hand gestures are typically used as commands to send information to the robot (e.g. move left, right, etc.). For example, Wachs et al. (2002) evaluated a recognition system using a fuzzy C-Means clustering method to classify a set of 12 static gestures. Results revealed an acceptance rate of 99.6% (percent of gestures with a sufficiently large membership value to belong to at least one of the designated classifications) and a recognition accuracy of 100% (the percent of accepted gestures classified correctly). The average recognition time was 8.43 ms., which allowed for real-time operation.

7.3.2.3 Evaluation of Handwriting Recognition

The evaluation of handwriting recognition systems has focused mainly on the recognition rate of symbols (characters, digits or words). For example, Yasuda et al. (2000) reported an average recognition rate of 85.35% in the recognition of several data sets containing Japanese Kanji (JIS first category), Katakana and Hiragana characters. In addition to recognition rate, other evaluation measures have been considered. For example, LeCun et al. (1995) focused on recognition time and memory requirements when evaluating several recognition algorithms (e.g. linear classifier, k-nearest neighbour classifier, pair wise linear classifier, etc.).

Several authors have reported on the success of handwriting recognisers used in MMDSs. For example, Vo and Wood (1996) reported on the Jeanie system, which was capable of processing writer-independent, continuous (cursive) handwriting recognition, achieving a recognition rate of over 90% on a 20,000 word vocabulary, whereas Johnston et al. (2002) reported a handwriting sentence accuracy of 64% in the evaluation of the MATCH system.

7.3.3 Evaluation of Multimodal Data Fusion

Several researchers have evaluated MMDSs focusing on the data fusion module, considering the combination of several input modalities, for example, speech and gestures, speech and visual information, and speech, gestures and visual information. The commonly used evaluation measure is the percentage of correct semantic interpretations obtained by the system due to the co-operation of the input modalities, although some users have also evaluated the recognition of the user emotional state.

In this section we address the evaluation of the fusion of speech and gestures, speech and visual information, and three modalities together.

7.3.3.1 Speech and Gestures

Vo and Wood (1996) evaluated the Jeanie system in understanding a small set of 185 user-system interactions: 77 speech only, 57 pen gesture only and 52 combination of both modalities. The results indicate that in absence of recognition errors of both recognisers, the fusion module obtained 80% correct semantic interpretations, which was acceptable for the system to be usable. However, if word accuracy was 76% the understanding rate decreased to 62% (the difference of 18% was due to 15% ASR errors and 3% gesture recognition errors).

Concerning the QuickSet system, Cohen et al. (1997) evaluated the fusion of pen and speech input considering additionally the effect of the user accent (native vs non-native English speakers), finding that the speech recognition rate was worse for accented users (-9.5%) although the gesture recognition was slightly higher ($+3.4\%$) for these users. The overall enhancement due to the multimodal interaction was higher for the accented users ($+15\%$) than for the native speakers ($+8.5\%$). The rate of multimodal recognition for accented users did not differentiate substantially from the results obtained for the native speakers. Also concerned with this system, Kaiser and Cohen (2002) trained a statistical model to assign weights to the gestural and speech recognisers for the task of processing 17 types of multimodal commands. On a test set of 759 commands, this model allowed to reduce the error rate for predicting the multimodal type of the 1st-best multimodal integration (i.e. the semantic interpretation to be used by the system) from 10.41% obtained in previous experiments to 4.74%, a relative reduction in error of 55.6%.

7.3.3.2 Speech and Visual Information

As discussed in Section 2.1.5, visual information extracted from the user face is very important for ASR, specially in situations where the auditory channel is degraded because

Table 7.6 Recognition results obtained from several acoustic conditions

Test type	Visual info. only (%)	Acoustic info. only (%)	Combined info. (%)
Clean data	55	98.4	99.5
SNR 16 dB	55	59.6	73.4
SNR 8 dB	55	36.2	66.5

of noise, bandwidth filtering or hearing-impairment. Some researchers have evaluated the fusion of both modalities focusing on perception studies. For example, early work by Risberg and Lubker (1978) found that subjects who watched a person speaking without hearing his voice perceived 1% of the words; if the subjects only heard the voice degraded by a low-pass filter perceived 6% of the words; and if they received both signals (visual and acoustic) perceived 45% of the words. Benoît et al. (1994) compared the perception of 18 nonsense words using only the auditory channel with that of using simultaneously the auditory and the visual channels, finding that the visual channel was almost unnecessary in clean acoustic conditions in which the SNR was greater than zero. However, they also observed that in very degraded acoustic conditions (SNR $= -24$ dB) seeing the user face allowed recognise 12 out of the 18 words. That degraded conditions are typical of discotheques, noisy streets and industrial plants in which the auditory channel alone does not allow recognising the words; thus, visual speech clearly increments the acoustic intelligibility.

Other authors have evaluated the fusion of both modalities considering the accuracy in recognising symbols. For example, Yang et al. (1998) evaluated an AVASR system trained with 170 sequences of acoustic/visual data in the task of recognising 30 sequences adding white noise. Table 7.6 sets out the results obtained, which show that the use of both types of information (visual and acoustic) clearly enhances the system performance, specially for accoustically degraded conditions (SNR 8 db).

Liu et al. (2002) used the same evaluation measure to evaluate the performance of an AVASR system in which the audio and visual information was integrated using a coupled hidden Markov model (CHMM). The experimental results, obtained on the XM2VTS database, showed that the error rate of the audio only ASR at SNR of 0 dB was reduced by over 55%.

Rogozan and Deléglise (1998) used the same measure to evaluate the recognition of sequences of letters spelled in French, using a method to adaptively integrate acoustic and visual information with different weights for each modality, depending on the current SNR. The results were obtained for different SNRs (-10, 0 and 10 dB) and different architectures for AVASR (direct identification, separate identification and hybrid identification, studied in Section 3.2.2.1). The results confirmed this adaptive weighting strategy allowed to enhance recognition accuracy. For example, in the case of the direct identification architecture, the accuracy for clean data was 88.3% in the case of equal weights, and 95.4% in the case of adaptive weights.

7.3.3.3 Speech, Gestures and Visual Information

Several researchers have focused on the fusion of speech, gestures and visual information to enhance the performance of MMDSs in several ways. For example, Adelhardt et al. (2003)

studied these input modalities to obtain information about the user emotional state (discussed in Section 3.2.5.1). They observed that this information can be useful to prevent the recurrent misunderstandings typical of dialogue systems, and proposed to combine emotional information obtained from the user voice, facial expressions and gestures. They relied on the idea that an angry voice may indicate a frustrated user, a joyful face may indicate a satisfied user, and a hesitant searching gesture of the user may reveal his unsureness. The facial expression recognition was based on eigenspaces (Yambor et al. 2000), the prosodic recognition was based on ANNs, and the gesture recognition was based on HMMs. The evaluation results (measured in terms of recognition accuracy) were 32%, 76% and 77% for facial expression, prosody and gesture recognition, respectively. According with the authors, these results shown the three input modalities could be used to identify the user emotional state; however, since a user state is not always indicated by the three modalities at the same time, a fusion of the different modalities seemed to be necessary (e.g. combining feature vectors of the three modalities to train an ANN, using a weighted sum of the recognition probabilities, etc.).

7.3.4 Evaluation of Animated Agents

The evaluation of animated agents is a very complex task. Experimental results reported by several authors show that the human appearance induces a more social behaviour on the part of the users (Bell and Gustafson 1999; Cerrato and Ekeklint 2002), which suggests the evaluation requires specialised methods. Evaluation standards are still lacking, partly due to the great complexity and variety of agents applied to a great variety of domains and applications (e.g., virtual guides, vendors, etc.). Given that these agents are used to increase the intelligibility of natural or synthetic speech as well as enhancing the interface visual appearance, it seems reasonable to think that their effectiveness must be measured in terms of their communicative abilities. Most evaluation measures use Likert scales on statements such as 'I trusted the agent', 'The agent was a kind person', 'The agent was bad-tempered', 'The movement of eyes, lips, head, and other part of the agent's body looked natural', etc.

In this section we report on evaluation methods that take into account the agent effects on the perceived system's effectiveness, naturalness and intelligibility, as well as on the user's comprehension and recall. Then, we focus on the effects on the user of the agent's facial expressions and its similarity to human behaviour. The chapter ends by discussing a methodological framework to evaluate these agents.

7.3.4.1 Effect on Effectiveness and Naturalness

A basic measure to evaluate animated agents is whether they really make the dialogue system more effective and natural in communicating with the user. In some applications this seems to be quite obvious while in others the question is not so easy to answer. To answer this question some researchers have focused on the communicative responses of users interacting with different types of animated agents. For example, Cassell and Thorisson (1999) studied three types of agent: some only provided propositional responses, other provided propositional responses as well as information about their emotional state, and the third type used both propositional and envelope (e.g. gaze and gesture) responses. The results indicated that the last type led to a more natural and efficient interaction and was the best ranked by test users on the basis of its linguistic ability.

7.3.4.2 Effect on Intelligibility

Other researchers have evaluated the increment of speech intelligibility when using animated agents, in comparison with the intelligibility of the acoustic stimuli only. For example, LeGoff et al. (1996) also carried out intelligibility tests of animated agents considering five levels of degradation due to acoustic noise. Intelligibility of voice-only was compared against a lip model, a facial model, and the face of the original speaker. The results confirmed the importance of the visual information on speech perception: the entire face provided two-thirds of the acoustic intelligibility when the acoustic transmission was degraded or lost, the facial model (with the tongue movements excluded) provided one half, and the lip model alone provided a third.

Using the same evaluation measure, Beskow (1997) evaluated two cartoon-like animated agents: Parke and Olga. In the experiments, two synthetic voices (male and female) as well as natural voice (male and female) were tested. Some 18 normal-hearing subjects participated in experiments carried out in eight different audiovisual conditions with audio degraded by noise at SNR 3dB. The results indicated that with the male synthetic voice the intelligibility score was 30% in the case of audio-only, 45% in the audiovisual case (Parke agent) and 47% in the same case (Olga agent). Using natural voice, intelligibility varied from 62% (audio-only) to 70% with the Parke agent (the Olga agent was not evaluated in this condition).

Karlsson et al. (2003) evaluated the intelligibility of sentences spoken in English, Danish and Swedish by a group of normal-hearing subjects and a group of hearing-impaired English subjects. The loss in audition for the English and Swedish impaired subjects was of 86 dB in average. The results showed that for the normal and the impaired users, the intelligibility with a synthetic face was significantly higher than for the audio only case: the average increment on intelligibility was 22% for both.

7.3.4.3 Effect on User Comprehension and Recall

Animated agents have also been assessed considering the effect they produce on user comprehension and recall. For example, Van Mulken et al. (1998) evaluated the effect of the PPP agent designed to comment on particular fragments of presentations and emphasise them with several types of gesture. For example, it could make gestures to express emotions (e.g. approval or disapproval), assist in the communication function (e.g. advise, recommend, dissuade), assist in referring expressions (e.g. looking an object and pointing at it), help to regulate the user-system interaction (e.g. looking at the user when referring to him), etc. The evaluation considered two variables: agent (present vs. absent) and information type (technical vs. non-technical). The results indicated the agent did not have any effect on understanding of presentations, and the information type did not influence this fact. The agent had a positive effect on the presentations, which were considered less difficult and more interesting. The information type affected the user opinion about the help provided by the agent: in the case of technical information, the agent was perceived as providing more help than in the case of non-technical information. In general terms, the agent was perceived positively: 50% of the users indicated that they would prefer presentations with an animated agent, independently of the presentation type, and 47% indicated that their preference would depend on the material of the presentation.

7.3.4.4 Effect of Facial Expressions

Several authors have evaluated animated agents paying attention to their facial expressions. For example, Koda and Maes (1996) studied the effect of using a face and facial expressions in a system to play poker, considering as evaluation measures the required attention, engagement and distraction. The authors also studied three aspects. The first was the type of facial features (e.g. age, humanity and realism) that made the agent look intelligent, desirable and comfortable to work with. The second was to know whether the user impressions about the agent were determined by its appearance, its performance or both. They also studied whether their judgements depend on gender or opinion about personification[7] (in favour or against) (Van Mulken et al. 1998; Koda and Maes 1996). The evaluation results, based on a 7-point Likert scale, indicate the users considered the poker-playing system equally intelligent independent of whether a face was used or not, which suggests that adding a face does not increase the perception about intelligence. However, having a face was considered as more desirable, engaging and comfortable to play with, independent of the user opinion about personification.

Animated agents have also been evaluated in terms of their success in transmitting specific emotions by their facial expressions. For example, Fabri et al. (2002) evaluated the use of a reduced set of FACS action units (Section 3.2.4.1) to generate the six basic emotions (surprise, fear, disgust, anger, happiness and sadness) in addition with a neutral expression. After randomly showing 28 images of animated agents to 29 subjects, a statistical analysis was carried out showing 62.2% average emotion recognition rate. Considering the same evaluation measure, Costantini et al. (2004) used MPEG-4 FAPs (also discussed in Section 3.2.4.1) obtained in two conditions: from real actors expressing different emotions (FAP condition) and from scripts specified by the developer (SB condition) to evaluate the expressiveness of two synthetic faces (face1 and face2). 14 videos showing three different faces (ACTOR, face1 and face2) were shown to 30 subjects (15 male, 15 female) who assigned to each facial expression an emotional label ('anger', 'happiness', 'neutral', 'disgust', 'surprise', 'fear' or 'sadness') depending on the perceived emotion. Table 7.7 shows the recognition rate for

Table 7.7 Percentage of correct recognition for each emotion and condition (reproduced by permission of Costantini et al. 2004)

	ACTOR1	F1-FAP	F1-SB	F2-FAP	F2-SB
Anger	90	27	53	7	23
Happiness	97	80	40	80	77
Neutral	70	70	60	53	67
Disgust	13	20	53	17	17
Surprise	47	40	87	33	90
Fear	50	17	77	0	77
Sadness	17	7	97	7	97
All	55	37	67	28	64

---

[7] The so-called 'personification' refers to the use of a human-like animated agent in the output interface of a MMDS.

each emotion and condition tested (the last row shows the average results per condition for all the emotions).

As can be observed, the average results obtained in the ACTOR condition were better than those obtained for both synthetic faces in the FAP condition, but not for that obtained in the SB condition. Also, the results for both synthetic faces were better in the SB condition than in the FAP condition.

7.3.4.5 Effects of Similarity with Human Behaviour

Animated agents have also been evaluated considering their similarity to human behaviour and the effect it produces on different evaluation measures. For example, Bickmore and Cassell (2004) evaluated the effects of an agent generating the small talk typical at the start of some human conversations. Two types of users were considered: introverts and extroverts. The authors also studied whether the non-verbal behaviour, which plays a very important role in human-human dialogues, plays the same critical role in human-system interaction. Users interacted with two versions of an animated agent differentiating only in that one only generated task-oriented interactions while the other featured in addition social dialogue to avoid threatening situations and increase trust. A subjective evaluation was carried out using a 9-point Likert scale on friendliness, credibility, life-like appearance, competence, reliability, efficiency, intelligence, etc. Subjects were also asked to give their opinion about to what extent they thought they got to know the agent, and to what degree the agent got to know and understand them, as well as how close they got to the agent. According to the authors, the experiments showed the users considered the interaction more like human-to-human than human-to-system, since their verbal lack of fluency was typical of a non-planned, conversational-style interaction. The results also indicated the social dialogue required less cognitive load that the task-oriented dialogue, the conversation required a greater cognitive load for introverts than extraverts, the dialogue on the telephone was more demanding than the face-to-face dialogue for extroverts, and the face-to-face dialogue was more demanding than the telephone dialogue for introverts.

Other human-like behaviours have been investigated by other researchers. For example, Heylen et al. (2002) compared a version of an animated agent that featured some patterns of human behaviour (e.g. signalling turn taking) with two other versions of the same system. In one of them the gaze movements were kept at a minimum, whereas in the other these movements were carried out randomly. The results indicated the users interacting with the system that implemented human patters appreciated the agent more, the interactions were more efficient and the users required less time to complete tasks.

7.3.4.6 A Methodological Framework

Buisine et al. (2002) proposed a methodological framework to guide the assessment of animated agents by separating a relevant set of variables that affect the evaluation into a set of hypotheses, as shown in Table 7.8.

The hypotheses are associated with variables that take into account the great diversity of animated agents. A variable is defined as an individual feature that has a possible influence on the evaluation, according to the experimenter hypotheses. To be testable, a variable must

Table 7.8 Proposed list of hypotheses to guide the evaluation of animated agents (reproduced by permission of Buisine et al. 2002)

	Hypothesis	To-be-measured criteria
H1	The use of an animated agent enhances the ergonomics of the interface.	Ergonomic criteria: guidance, workload, explicit control, adaptability, error management, consistency, significance of codes, compatibility.
H2	The use of an animated agent enhances the effectiveness of the interaction.	Speed of interaction (time to achieve the goal, navigation), performance or achievement of intermediate goals (number of errors).
H3	The use of an animated agent enhances the satisfaction of the user.	Self-rated pleasantness, effectiveness, usefulness, case of use, ease to learn ...
H4	Multimodal behaviour of users depends on multimodal behaviour of animated agents.	Gesture displayed, types of multimodal cooperation.

Table 7.9 Proposed list of variables to manipulate in the evaluation of animated agents (reproduced by permission of Buisine et al. 2002)

	Variables	Values	Remarks
V1	use of verbal communication	in input; in output; both; none	
V2	additional vocal cues	True/false	e.g. intonation, intensity, pitch, tone ...
V3	additional nonverbal communication	in input; in output; both; none	
V4	type of nonverbal cues	gesture; facial expression; both; none	
V5	type of multimodal cooperation	Equivalence; specialization; transfer; redundancy; complementarity; concurrency; none	
V6	agent's exhibited abilities and skills	Ability to engage a conversation; expression of personality; expression of emotions; none	several values can coexist within a single agent.
V7	amount of embodiment	Face-only; full-body	
V8	style of rendering	2D; 3D	
V9	realism	Cartoonist; photo-realistic	
V10	sophistication of animation	Still images; cartoon; realistic	
V11	type of voice in output	synthetic; natural	
V12	fit of the agent with the user's characteristics	True/false	e.g. same age, same sex
V13	fit of the agent with the user's preferences	True/false	e.g. opposite sex, funny characters for children ...
V14	fit of the agent with the task characteristics	True/false	e.g. air hostess for a flight reservation service ...

comprise several values (at least two: presence vs. absence of the feature). Table 7.9 shows the variables the authors proposed to guide the evaluation, some of which have been previously investigated by other authors; for example, V1 was studied by Caelen and Bruandet (2001), V3 by Granström et al. (2002) and Moreno et al. (2001), V6 by Bickmore and Cassell (2001) and Cassell et al. (1999); V7 was studied by McBreen and Jack (2001), V8 by Traum and Rickel (2001), and V9 by Koda and Maes (1996) and Traum and Rickel (2001).

The framework relies on using the values of the variables to create experimental groups in which to study the effects of such variables. All the groups must have the same values for the non-studied variables. The aim is to observe the effect of the studied variable, using statistical measures of the user opinions about a set of evaluation measures. The authors applied this framework to observe the effect of two variables (multimodal dialogue strategy of an animated agent, and agent appearance) on several subjective measures (explanation quality, trust, likeability, personality and expressivity), considering as well a performance measure (user recall). The evaluation subjects were two groups of users (9 male, 9 female) who observed three short presentations displayed together with an animated agent. The presentations were about a video-editing software, a video projector remote control, and a copy machine. Three types of multimodal strategies were evaluated (i.e. this variable had three values): cooperation of modalities by redundancy (the relevant information was provided to the user using speech and deictic gestures), cooperation by complementarity (one half of the information was provided by each modality) and cooperation by specialisation (all the information was provided by speech). To evaluate the effect of the variable 'agent appearance', three different 2D animated agents were used. After the presentations, test users were given three graphics regarding the presentations they previously saw in order to determine the amount of information they were able to recall, which was marked 0–10 by an experimenter. The subjects filled out a questionnaire giving their opinions about the three animated agents in terms of the subjective evaluation measures considered. The evaluation results indicated, on the one hand, the multimodal strategy featured by the animated agent influenced the subjective opinion about explanation quality, but not that about trust, likeability, personality, expressivity and recall. On the other, they showed the agent appearance had a notable effect on likeability and recall, but not on explanation quality and trust.

7.4 Summary

In this chapter we studied evaluation techniques for SDSs, multilingual and MMDSs. Initially we classified these techniques into four categories: subsystem-level, end-to-end, dialogue processing and system-to-system. The study began with a discussion of the evaluation of spoken and multilingual dialogue systems. We first considered the subsystem-level evaluation, focused on the ASR, NLU, NLG and speech synthesis modules. Next, we addressed the end-to-end and dialogue processing evaluation. In the latter case, we studied the PARADISE framework as well as an analytical approach. In the following section of the chapter we focused on the system-to-system evaluation, addressing the evaluation with limited resources, linguistic noise and a user simulation technique. In the third section we addressed the evaluation of MMDSs. Initially, we focused on the system-level evaluation, discussing three approaches (experimental, predictive and expert-based) and commented on the PROMISE evaluation framework. In the next section we addressed the evaluation at the subsystem level. We first addressed the evaluation of face localisation and gaze tracking, gesture recognition and handwriting recognition modules. Then we addressed the evaluation of multimodal data fusion module, focusing on the fusion of speech and gestures, speech and visual information, and finally, the three modalities together. To conclude, we addressed the evaluation of animated agents, focusing on the effects they produce on the system's effectiveness and naturalness, intelligibility, and on the user's comprehension and recall. Then we studied the effects on the user of the agent's facial expressions and its similarity with the human behaviour. Finally, we studied a methodological framework to guide the evaluation of animated agents.

7.5 Further Reading

- Section 2.3 of Gibbon et al. (2000).
- Proc. of Workshop 'Multimodal Resources and Multimodal Systems Evaluation', Language Resources and Evaluation Conference, 2002.

Appendix A

Basic Tutorial on VoiceXML

As discussed in Section 6.1.2.1, VoiceXML is an XML-based, markup language that eases the development of voice applications (SDSs) combining ASR, DTMF, audio recording, speech synthesis, playback of audio files, HTTP requests, telephone call transfer, etc. The conceptual design of a VoiceXML processor can be described as shown in Figure A.1.

The data the user is willing to achieve (e.g. train departure schedule) is requested by the user's input (speech or DTMF) and delivered back via TTS or recorded audio. This data is included in the system's spoken responses, which can be prepared in advance or generated dynamically at run-time.

Although VoiceXML was originally designed for developing simple IVR (Interactive Voice Response) systems, it can be used for developing flexible SDSs for several applications, due to its flexible mechanism for dialogue control and the availability of high level programming elements such as *if-then-else*, scope of variables, handling of thrown events, subdialogue calls, etc.

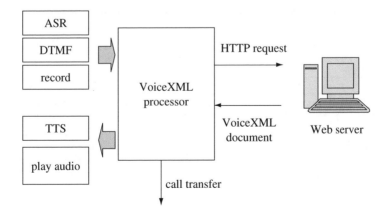

Figure A.1 Conceptual design of a VoiceXML processor

Spoken, Multilingual and Multimodal Dialogue Systems: Development and Assessment Ramón López-Cózar Delgado
and Masahiro Araki © 2005 John Wiley & Sons, Ltd

Basic Elements of VoiceXML

VoiceXML has a hierarchical structure, with a root element and some child elements. The root element is `<vxml>` which has two required attributes: *version* and *xmlns* (XML namespace of VoiceXML). An application typically consists of a set of documents, which can take a flat structure (like a set of HTML documents) or a two-level structure. In the second case (Figure A.2), at the higher level the application root document has a set of variables and grammars which are shared with its low-level child (or *leaf*) documents. In the child elements, the root document is indicated via URI (Universal Resource Identifier) in the *application* attribute of the `<vxml>` element.

As a child element of `<vxml>`, there are two types of basic dialogues: `<form>` and `<menu>`. The `<form>` element, which is conceptually equal to the `<form>` tag in HTML, represents a set of variables which are treated within one dialogue segment. The `<form>`

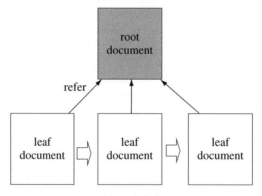

Figure A.2   Two-level document structure of VoiceXML

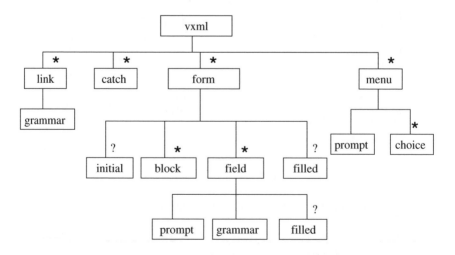

Figure A.3   Basic structure of VoiceXML tags

element has <field> elements for obtaining values by the interaction with the user. Each <field> element may include a <prompt> element (a system's utterance as for example 'Please enter the departure station') to obtain a particular data (e.g. *departureStation*), a <grammar> to specify the set of possible user sentences that can be recognised in response to the prompt, and control information represented by <block> and <filled> elements, which contain executable content. The <menu> element provides a simplified representation of dialogue branches, containing a <prompt> element and a set of <choice> elements that specify a keyword for each branch and a target URI.

The basic structure of VoiceXML tags (also called *elements*) is shown in Figure A.3, where '*' means the element can appear zero or more times and '?' means it can appear once or zero times.

System-Initiative Dialogue

In order to explain the basic elements of VoiceXML, we describe in this section the construction of a simple dialogue system that features system-directed interaction to provide conference information (e.g. outline of the conference, access information and registration). The document structure (*site map*) for this example is shown in Figure A.4.

Readers familiar with HTML and Web technologies can easily imagine what this site will look like in terms of HTML pages. Using VoiceXML, these pages are replaced by VoiceXML documents which convert the GUI into a voice-enabled site.

Menu dialogue

In Figure A.4, the *Top page* (*document* in VoiceXML terminology) is the first document loaded in the VoiceXML processor when the system catches the user call. Its goal is to introduce the site and guide the user to the desired service. Such a guide can be implemented using a <menu> element as shown in Figure A.5.

When this page is loaded on the VoiceXML processor, the first child of the <vxml> element (i.e. the <menu> element) is executed. In the <menu> element, the <prompt> element specifies what the system says. The content text in the <prompt> element is passed

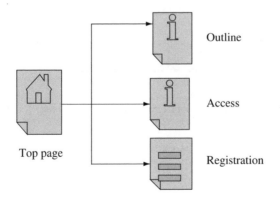

Figure A.4 Site map for a conference information application

```
<?xml version="1.0" ?>
<vxml version="2.0" xml:lang="en-US"
  xmlns="http://www.w3.org/2001/vxml">
 <menu>
  <prompt>
     This is the conference guide voice service.
     Chose one of the following, <enumerate/>
  </prompt>
  <choice next="outline.vxml"> outline </choice>
  <choice next="access.vxml"> access </choice>
  <choice next="register.vxml"> registration </choice>
 </menu>
</vxml>
```

Figure A.5   Top VoiceXML document (top.vxml)

to the TTS engine to be transformed into the system's voice. SSML (Speech Synthesis Markup Language) tags can be added if the developer wants to add prosodic features to the voice output. The `<enumerate>` element is replaced by the contents of the set of `<choice>` elements (i.e. 'outline', 'access' and 'registration'). The `<choice>` element specifies the keyword for the branch and the target element as the *next* attribute.

Dialogue for Providing Information to the User

The dialogue for providing information about the conference to the user is shown in Figure A.6.

As can be observed, there is only one `<block>` element in the `<form>` element. Therefore, this form element does not gather information from the user. It only provides the information specified in the contents of the `<block>` element and returns to the menu choice dialogue (Figure A.5) as specified by the `<goto>` element.

Dialogue to Collect Data from the User

The dialogue to collect information from the user is shown in Figure A.7. It gathers the necessary data to register the user at the conference.

```
<?xml version="1.0" ?>
<vxml version="2.0" xml:lang="en-US"
  xmlns="http://www.w3.org/2001/vxml">
 <form>
  <block>
     The conference will take place at Kyoto Conference
       Center, April 1, 2005.
     The theme is...
   <goto next="top.vxml"/>
  </block>
 </form>
</vxml>
```

Figure A.6   VoiceXML document for providing information (outline.vxml)

```
<?xml version="1.0" ?>
<vxml version="2.0" xml:lang="en-US" xmlns="http://www.w3.org/2001/vxml">
  <form>
    <block> This is the registration section.</block>
    <field name="participant">
     <prompt> What's your name? </prompt>
     <grammar src="name.grxml" type="application/srgs+xml"/>
    </field>
    <field name="member_id" type="digits">
     <prompt> What's your membership number? </prompt>
    </field>
    <field name="reception" type="boolean">
     <prompt> Are you going to attend the reception? </prompt>
    </field>
    <filled>
     <submit next="http://localhost:8080/register"/>
    </filled>
  </form>
</vxml>
```

Figure A.7   VoiceXML document to collect data from the user (register.vxml)

As can be observed, the first <block> indicates the service provided by this document. The following three <field> elements correspond with the variables 'participant', 'member_id' and 'reception', respectively. Each <field> specifies the name of the variable by a *name* attribute, the system's utterance (<prompt>) to obtain a value for the variable, and the set of possible user responses that can be recognised by the <grammar> element (or by the *type* attribute of the <field> element). If the expected user sentences match a built-in grammar such as date, currency, phone, etc., the pre-defined grammar can be used, and thus specified by the *type* attribute of the <field> element. Otherwise, the developer must define a grammar to obtain the use input, which is specified by the <grammar> element. Such a grammar can be *in-line* (delimited by the <grammar> ... </grammar> elements) or in a external file, which is indicated by the *src* attribute. The recommended grammar format is SRGS (Speech Recognition Grammar Specification) in either ABNF (Augmented BNF) or XML format.

After all the field variables have been assigned a value from the user's input, the content of the <filled> element is executed. Typically, operations carried out in the execution are (1) consistency check of the acquired data (e.g. the 'member_id' is below the max number) using <if>, <then>, <else> elements; and (2) forwarding of the data to a server-side program by using the <submit> element, with its *next* attribute specifying the corresponding URI.

By default, the dialogue proceeds sequentially along with the sequence of child elements of the <form> element, in accordance with the Form Interpretation Algorithm (FIA). This algorithm decides the next form item to visit, and considers that all the form items, such as <field>, <block>, etc., have associated a variable either explicitly declared or *hidden*, which is initially set to 'undefined'. In order to decide the next item to visit, the FIA searches the form in a top-down fashion to find the first item whose variable is 'undefined', and then visits (i.e. executes) this item. For example, if the selected item is a <field> element, the dialogue system will try to collect a value for this field. When the item variable is assigned a value, it cannot be visited again by the FIA.

```
<?xml version="1.0" ?>
<grammar version="1.0" xmlns=http://www.w3.org/2001/06/grammar
    tag-format="semantics/1.0" xml:lang="en-US"
    mode="voice" root="main_rule">
 <rule id="main_rule" scope="public">
    <ruleref uri="#first" /> <ruleref uri="#last" />
 </rule>
 <rule id="first">
    <one-of>
      <item> John </item>
      <item> Bill </item>
      ...
    </one-of>
  </rule>
 <rule id="last">
    <one-of>
      <item> Doe </item>
      <item> Gates </item>
      ...
    </one-of>
  </rule>
</grammar>
```

Figure A.8   SRGS representation (name.grxml)

Grammar Format

According with the VoiceXML 2.0 specification, all the platforms must support SRGS. For example, a SRGS representation for the participant's name can be as shown in Figure A.8.

A `<rule>` element corresponds to a rule of a Context Free Grammar (CFG). The value of the *id* attribute corresponds to the left-hand-side symbol of the CFG rule. A `<ruleref>` element corresponds to one non-terminal symbol of the right-hand-side of the CFG rule. If the content of a `<rule>` element is a `<one-of>` element, it means a *word dictionary*. The word, i.e. terminal symbol, is specified by the non-XML contents of the `<item>` element. This `<item>` element also can be used as a *repeat* marker, like the symbols '+', '*', '?' in typical regular expressions. As a value of the *repeat* attribute of the `<item>` element, the symbols '+', '*', '?' can be represented as '1−', '0−', '0−1', respectively.

Example Dialogue

Using the VoiceXML documents commented above, an example dialogue can be as shown in Figure A.9 (where 'S' stands for dialogue system and 'U' for user).

Mixed-Initiative Dialogue

In this section we explain two types of mixed-initiative dialogue patterns. In the first one the system jumps to another contents regardless of the current dialogue state (*unexpected initiative taking*), while in the second it converses with the user exchanging the initiative in the dialogue (*expected initiative taking*).

S: This is the conference guide voice service.
  Choose one of the following, outline, access, registration.
U: Outline.
S: The conference will take place at Kyoto Conference Center,
  April 1, 2005. The theme is...
  This is conference guide voice service.
  Choose one of the following, outline, access, registration.
U: Registration.
S: This is the registration section.
  What's your name?
U: John Doe.
S: What's your membership number?
U: 12345.
S: Are you going to attend the reception?
U: yes.
  (*jumps to server-side registration program*)

Figure A.9   Example of user-system dialogue using VoiceXML

Unexpected initiative taking

Generally speaking, dealing with the user unexpected initiative taking is very difficult because the system cannot be prepared for when it happens and for what the user says. VoiceXML has a very limited function to deal with this user behaviour: simply jumping to another dialogue. This provides a functionally which is equivalent to the frame representation in a GUI-based web page, in which a navigator window shows a menu that does not change as the user is visiting the site. This menu allows the user jump to another page at anytime. Similarly, VoiceXML can have an ASR grammar active that captures the user jumping utterances. In accordance with the VoiceXML scope mechanism shown in Figure A.10, this *anytime activate* grammar must be defined in the application root document in order to be accessible from any document of the application.

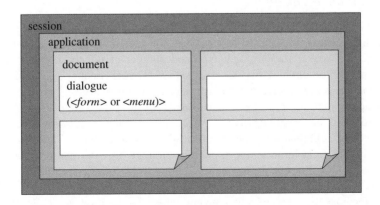

Figure A.10   VoiceXML scope

```
<?xml version="1.0" ?>
<vxml version="2.0" xml:lang="en-US"
      xmlns="http://www.w3.org/2001/vxml">
 <link next="outline.vxml">
  <grammar> outline </grammar>
 </link>
 <link next="access.vxml">
  <grammar> access </grammar>
 </link>
 <link next="register.vxml">
  <grammar> register </grammar>
 </link>
 <menu>
  (in the same contents of Figure A.5)
```

Figure A.11   Application root document (top.vxml)

In this figure, the session level scope manages read-only variables mainly concerned with the telephone session. The application level scope is carried out in the two-level application structure shown in Figure A.2. The variables, grammars and event handlers (e.g. to cope with user's help requests) can be accessed within the same application. The document level scope and the dialogue level scope can define their own local variables, grammars and event handlers.

To handle the unexpected jump taking within this scope mechanism, the application root document must have at least a `<link>` element that captures jumping commands by a `<grammar>` element. For example, in order to change the voice site described so far to an *anytime-jump* site, the top page must take the role of the application root document, and include a set of `<link>` elements as shown in Figure A.11.

In order to handle the *anytime jumping* to another document in the application, each child document (Figures A.6 and A.7) must indicate the same URI pointing to the application root document, using the *application* attribute of the `<vxml>` element.

Expected Initiative Taking

Finally, we consider another application type suitable for mixed-initiative dialogue: a railway ticket vending machine. In order to buy a ticket, the user must provide information about the departure and arrival stations, seat type and number of tickets. For novice users, an appropriate interaction strategy may be a *system-directed* (Section 5.3.1), with the system asking for a single piece of data in each turn (e.g. departure station). However, this strategy is surely inappropriate for expert users who may consider the system too slow and very inflexible. To avoid this negative effect, the system can employ a *mixed-initiative* interaction strategy (Section 5.3.1) that gives the initiative initially to the user and follows the dialogue according with the input information. Using this strategy, the user can provide all the data items required to buy a ticket in just one turn. Moreover, if necessary, the system can take the initiative to collect a particular data item from the user. In order to set up this strategy, VoiceXML includes the `<initial>` element, which can be used as a child of a `<form>` element as observed in Figure A.12.

```
<?xml version="1.0" ?>
<vxml version="2.0" xml:lang="en-US"
      xmlns="http://www.w3.org/2001/vxml">
  <form>
      <grammar src="ticket.grxml" type="application/srgs+xml"/>
      <initial name="ticket">
          <prompt> Ticket service. May I help you? </prompt>
      </initial>
      <field name="from">
          <prompt> Please say the departure station. </prompt>
            <grammar src="station.grxml" type="application/srgs+xml"/>
      </field>
      <field name="to">
          <prompt> Please say the arrival station. </prompt>
            <grammar src="station.grxml" type="application/srgs+xml"/>
      </field>
      <field name="number" type="number">
          <prompt> How many tickets do you want? </prompt>
      </field>
      <field name="seat" >
          <prompt> Seat type? Please say business or coach </prompt>
          <option> business </option>
          <option> coach  </option>
      </field>
      <filled>
              <submit next="http://localhost:8080/ticket"/>
      </filled>
  </form>
</vxml>
```

Figure A.12   Mixed-initiative interaction in VoiceXML

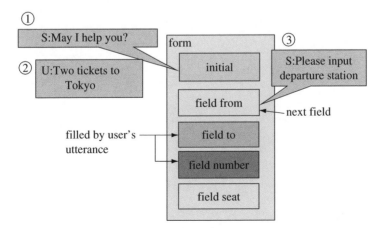

Figure A.13   FIA performance in mixed initiative dialogue

The form level grammar 'ticket.grxml' is defined to accept any combination of data (e.g. 'From Kyoto to Tokyo', 'Two business to Tokyo', etc.). Following the FIA algorithm explained above, the first form item to be visited is the `<initial>` element. After giving the initiative to the user by means of the prompt 'May I help you?', the system expects to fill in one or more form fields from the user input. For example, after processing the utterance 'Two tickets to Tokyo', the first unfilled `<field>` is 'from'. Thus, the FIA visits this field using the system-initiative strategy. The dialogue flow is represented graphically in Figure A.13.

Appendix B

Multimodal Databases

Database	Database description
	Faces, face recognition, facial expressions and eye movement
AR Face Database	Face recognition and facial expression recognition http://www-prima.inrialpes.fr/FGnet/data/05-ARFace/tarfd_markup.html
AT&T (Olivetti)	40 subjects, 10 images per subject
Cohn-Kanade	Facial expression AU-coded facial expression database. Development and test of algorithms for facial expression recognition
CMU PIE database	Pose, illumination and expression. Design and evaluation of face recognition algorithms and facial expression detection
EMED	Eye movement equipment database http://ibs.derby.ac.uk/emed
FERET	Male and female faces. Each contains a single person with certain expression http://nist.gov/humanoid/feret
Harvard	Cropped, masked face images under a wide range of lighting conditions ftp://ftp.hrl.harvard.edu/pub/faces
JAFEE	Facial expression
M2VTS	Multimodal database containing various image sequences http://poseidon.csd.auth.gr/M2VTS/index.html
MIT	Faces of 16 people, 27 of each person, various illumination conditions, scale and head orientation ftp://whitechapel.media.mit.edu/pub/images

(*continued overleaf*)

Spoken, Multilingual and Multimodal Dialogue Systems: Development and Assessment   Ramón López-Cózar Delgado and Masahiro Araki   © 2005 John Wiley & Sons, Ltd

(continued)

Database	Database description
Purdue AR	3,276 face images with different facial expressions and occlusions under different illuminations http://rvl1.ecn.purdue.edu/~aleix/ aleix_face_DB.html
UMIST	564 images of 20 subjects. Each subject covers a range of poses from profile to frontal views http://images.ee.umist.ac.uk/danny/ database.html
University of Bern	300 frontal face images of 30 people ftp://iamftp.unibe.ch/pub/Images/ FaceImages/
Yale	Face images with expressions, glasses under different illumination conditions http://cvc.yale.edu

Audiovisual

DAVID	Digital Audio-Visual Integrated database Data about 100 persons regarding speech or person recognition, synthesis of animated agents, voice control of video-conferencing resources
TULIPS 1.0	Lip tracking http://mplab.ucsd.edu/databases/databases.html#tulips

Face tracking, lip tracking nose and eyes

LIMSI Gaze corpus (CAPRE)	Track face, nose and eyes http://www.limsi.fr/Individu/collet/Public_html/ CapRe.eng.html Facial expression and gesture
Scan MMC (Score Analysed Multimodal Communication)	Research on facial expressions and gestures

Gesture

ATR sign language gesture corpora	Most important words of Japanese sign language as a basis for the development and evaluation of gesture recognition systems
LIMSI Pointing Gesture Corpus (PoG)	Basis for the specification of a recognition system

Gestures and audio

ATR multimodal human-human interaction database	Source for analysing the relation between speech and gesture

(continued)

Database	Database description
CHCC OGI	Multimodal real state map http://www.cse.ogi.edu/CHCC/index.html
GRC multimodal dialogue	Study patterns of multimodal communication during a work session about collaborative conception
IRISA Georal multimodal corpus	Speech and gestures made on a touch-sensitive screen showing a tourist map
LORIA Multimodal dialogue corpus	Research
RWC multimodal database of gestures and speech	Speech and video database to promote research and development of multimodal interactive systems integrating speech and video data
	Gestures, gaze and audio
LIMSI	Multimodal dialogues between car driver and co-pilot
VISLab	Cross-modal analysis of signal and sense data and computational resources for gesture, speech and gaze research http://vislab.cs.vt.edu/

Appendix C

Coding Schemes for Multimodal Resources

Intended for markup of	Name of coding scheme	Purpose of creation
Gaze	The alphabet of eyes	Analyse any single item of gaze in videotaped data.
Facial expression	FACS	Encode facial expressions by breaking them down into component movements of individual facial muscles (Action Units). Suitable for video or image.
	BABYFACS	Based on FACS but tailored to infants.
	MAX (Maximally Discriminative Facial Movement Coding System)	Measure emotion signals in the facial behaviour of infants and young children. Suitable for video or image.
	MPEG-4	Define a set of parameters to define and control facial models.
	ToonFace	Code facial expression with limited detail. Developed for easy creation of 2D synthetic interface agents.
Gesture	HamNoSys	Designed as a transcription scheme for (different) sing languages.

(*continued overleaf*)

Spoken, Multilingual and Multimodal Dialogue Systems: Development and Assessment   Ramón López-Cózar Delgado
and Masahiro Araki   © 2005 John Wiley & Sons, Ltd

(continued)

Intended for markup of	Name of coding scheme	Purpose of creation
	SWML (Sing Writing Markup Language) MPI GesturePhone MPI Movement Phase Coding Scheme	Code utterances in sign languages written in the Sign Writing System. Transcribe signs and gestures. Coding of co-speech gestures and signs.
Speech and gesture	DIME (Multimodal extension of DAMSL)	Code multimodal behaviour (speech and mouse) observed in simulated sessions in order to specify a multimodal information system.
	HIAT (Halbinterpretative Arbeitstranskriptionen)	Describe and annotate parallel tracks of verbal and non-verbal (e.g. gestural) communication in a simple way.
	TYCOON	Annotation of available referable objects and references to such objects in each modality.
Text and gesture	TUSNELDA	Annotation of text and image sequences, e.g. from comic strips.
Speech, gesture, gaze	LIMSI Coding scheme for multimodal dialogues between car driver and copilot	Annotation of a resource which contains multimodal dialogue between drivers and copilots during real car driving tasks. Speech, hand gesture, head gesture, gaze.
Speech, gesture and body movement	MPML (A Multimodal Presentation Markup Language with Character Agent Control Functions)	Allow users to encode the voice and animation of an agent guiding a web site visitor through a web site.
Speech, gesture, facial expression	SmartKom Coding Scheme	Provide information about the intentional information contained in a gesture.

Reproduced by permission of Dybkjær and Bernsen (2002)

Appendix D

URLs of Interest

ACL	http://www.aclweb.org
Apple	http://www.apple.com/macos/speech
AT&T	http://www.att.com/
BBN	http://www.bbn.com/
BeVocal Café	http://cafe.bevocal.com
C-STAR	http://www.c-star.org/
CCXML	http://www.w3.org/TR/ccxml/
CMU-Cambridge Statistical LM Toolkit	http://svr-www.eng.cam.ac.uk/~prc14/toolkit.html
CSLU toolkit	http://cslu.ece.ogi.edu/toolkit/
EMMA	http://www.w3.org/TR/emma/
Festival	http://www.cstr.ed.ac.uk/projects/festival/
HTK	http://htk.eng.cam.ac.uk/
IBM	http://www-306.ibm.com/software/voice/
ISCA	http://www.isca-speech.org/
ISTC	http://www.astem.or.jp/istc/index_e.html
Lernout & Hauspie	http://www.lhs.com/
Microsoft	http://www.microsoft/speech
Multimodal Interaction Activity	http://www.w3.org/2002/mmi/
Nuance	http://www.nuance.com
Philips	http://www.speechrecognition.philips.com/
ScanSoft	http://www.scansoft.com/
Speech Recognition Grammar Specification	http://www.w3.org/TR/speech-grammar/
Speech Synthesis Markup Specification	http://www.w3.org/TR/speech-synthesis/
Tellme Studio	http://studio.tellme.com/
TRINDI	http://www.ling.gu.se/projekt/trindi/trindikit/

Spoken, Multilingual and Multimodal Dialogue Systems: Development and Assessment   Ramón López-Cózar Delgado
and Masahiro Araki   © 2005 John Wiley & Sons, Ltd

VHML	http://www.vhml.org
Voice Browser Activity	http://www.w3.org/Voice/
VoiceXML	http://www.w3.org/TR/voicexml20/
VoiceXML Forum	http://www.voicexml.org/
VoiceGenie Developer workshop	http://developer.voicegenie.com
Voxeo Community	http://community.voxeo.com
Xface	http://xface.itc.it
XHTML + Voice profile	http://www.voicexml.org/specs/multimodal/x + v/12/

Appendix E

List of Abbreviations

ADM	Adaptive Delta Modulation
ADPCM	Adaptive Differential Pulse Code Modulation
ANN	Artificial Neural Network
APML	Affective Presentation Markup Language
ASR	Automatic Speech Recognition
AU	Action Unit
AVASR	Audio-Visual Automatic Speech Recognition
ATIS	Air Travel Information Service
BNF	Backus Naur Form
CA	Concept Accuracy
CHMM	Coupled Hidden Markov Model
CMU	Carnegie Mellon University
DM	Delta Modulation
DMML	Dialogue Manager Markup Language
DPCM	Differential Pulse Code Modulation
DPML	Discourse Plan Markup Language
DSP	Digital Signal Processing
DTMF	Dual Tone Multiple Frequency
EMMA	Extensible Multimodal Annotation Markup Language
FACS	Facial Action Coding System
FAP	Facial Animation Parameter
FDP	Facial Definition Parameter
FFT	Fast Fourier Transformation
FIA	Form Interpretation Algorithm
FP	Feature Point
GUI	Graphic User Interface
HCI	Human-Computer Interaction
HML	Human Markup Language
HMM	Hidden Markov Model
HTK	Hidden Markov Model Toolkit

Spoken, Multilingual and Multimodal Dialogue Systems: Development and Assessment Ramón López-Cózar Delgado and Masahiro Araki © 2005 John Wiley & Sons, Ltd

HTTP	Hyper Text Transfer Protocol
ICL	Interagent Communication Language
IPA	International Phonetic Alphabet
JSGF	Java Speech Grammar Format
JSP	Java Server Pages
LCD	Liquid Crystal Display
M3L	Multimodal Markup Language
MIT	Massachusetts Institute of Technology
MLDS	Multilingual Dialogue System
MMDS	Multimodal Dialogue System
MMIL	Multimodal Interface Language
MPML	Multimodal Presentation Markup Language
MS-TDNN	Multi-State Time Delay Neural Networks
MT	Machine Translation
MURML	Multimodal Utterance Representation Markup Language
MVC	Model-View-Controller
N.A.	Not Applicable
NLG	Natural Language Generation
NLP	Natural Language Processing
NLU	Natural Language Understanding
OAA	Open Agent Architecture
OCR	Optical Character Reading
OWL	Web Ontology Language
PCM	Pulse Code Modulation
PDA	Personal Digital Assistant
PLS	Pronunciation Lexicon Specification
POS	Part-of-Speech
RDF	Resource Description Framework
ROI	Region of Interest
SA	Sentence Accuracy
SALT	Speech Application Language Tags
SDK	Software Development Kit
SDS	Spoken Dialogue System
SMIL	Synchronised Multimedia Integration Language
SNR	Signal-to-Noise Ratio
SQL	Structured Query Language
SRGS	Speech Recognition Grammar Specification
SSML	Speech Synthesis Markup Language
TDNN	Time Delay Neural Network
TFS	Typed Feature Structure
TTS	Text-To-Speech
WER	Word Error Rate
WOz	Wizard of Oz
W3C	World Wide Web Consortium
XHTML	eXtensible HyperText Markup Language
XML	Extensible Markup Language

References

Abbott, K. R. 2002. *Voice Enabling Web Applications: VoiceXML and Beyond.* a! Press.

Abry, C., Lallouache, M. T. 1991. Audibility and stability of articulatory movements: Deciphering two experiments on anticipatory rounding in French. *Proc. 12th International Congress of Phon. Sci.*, Aix-en-Provence, France, pp. 220–225.

Adelhardt, J., Shi, R., Frank, C., Zeissler, V., Batliner, A., Nöth, E., Niemann, H. 2003. Multimodal user state recognition in a modern dialogue system. *Proc. 26th German Conference on Artificial Intelligence*, pp. 591–605.

Alexandersson, J., Becker, T. 2001. Overlay as the basic operation for discourse processing in a multimodal dialogue system. *Proc. 2nd IJCAI Workshop on Knowledge and Reasoning in Practical Dialogue Systems*.

Alexandersson, J., Becker, T. 2003. The formal foundations underlying overlay. *Proc. 5th International Workshop on Computational Semantics*, pp. 22–36.

Allen, J. 1995. *Natural language understanding.* The Benjamin/Cummings Publishing Company Inc.

Allen, J. F., Ferguson, G., Miller, B., Ringger, E. 1995. Trains as an embodied natural language system. *Proc. AAAI Symposium on Embodied Language and Action*, Technical Report FS-95-05.

Allen, J. F., Miller, B. W., Ringeer, E. K., Sikorski, T. 1996. A robust system for natural spoken dialogue. *Proc. 34 Annual Meeting of the ACL*, pp. 62–70.

Almeida, L., Amdal, I., Beires, N., Boualem, M., Boves, L., den Ons, E., Filoche, P., Gomes, R., Knudse, J. E., Kvale, K., Rugelbak, J., Tallec, C., Warakagoda, N. 2002. Implementing and evaluating a multimodal and multilingual tourist guide. *Proc. ISCA Tutorial and Research Workshop on Multimodal Dialogue in Mobile Environments*.

André, E., Klesen, M., Gebhard, P., Allen, S., Rist, T. 1999. Integrating models of personality and emotions into lifelike characters. *Proc. International Workshop on Affect on Interactions – Towards a New Generation of Interfaces*, pp. 136–149.

André, E., Müller, J., Rist, T. 1996. The PPP Persona: A multipurpose animated presentation agent. *Proc. Advanced Visual Interfaces*, ACM Press.

Ang, J., Dhillon, R., Krupski, A., Shriberg, E., Stolcke, A. 2002. Prosody-based automatic detection of annoyance and frustration in human-computer dialog. *Proc. ICSLP*, pp. 2037–2040.

Antoniou, G., Harmelen, F. V., Antoniou, G. 2004. *A Semantic Web Primer*, MIT Press.

Araki, M., Doshita, S. 1997. Automatic evaluation environment for spoken dialogue systems. In E. Mayer, M. Mast, and S. LuperFoy (eds), *Dialogue Processing in Spoken Language Systems*, Springer, pp. 183–194.

Araki, M., Komatani, K., Hirata, T. and Doshita, S. 1999. A dialogue library for task-oriented spoken dialogue systems. *Proc. IJCAI Workshop on Knowledge and Reasoning in Practical Dialogue Systems*, pp. 1–7.

Aust, H., Oerder, M. 1995. Dialogue control in automatic inquiry systems. *ESCA Workshop on Spoken Dialogue Systems*, pp. 121–124.

Aust, H., Oerder, M., Seide, F., Steinbiss, V. 1995. The Philips automatic train timetable information system. *Speech Communication*, 17: 249–262.

Balci, K. 2004. Xface: MPEG-4 based open source toolkit for 3D facial animation. *Proc. Working Conference on Advanced Visual Interfaces*, pp. 399–402.

Balci, K. 2005. Xface: Open source toolkit for creating embodied conversational agents. *Proc. SmartGraphics Conference*, Lecture Notes in Computer Science (*forthcoming*).

Ball, G., Breese, J. 2000. Emotion and personality in a conversational character. In Cassell, J., Sullivan, J., Prevost, S., Churchill, E. (eds.) *Embodied Conversational Agents*, MIT Press.

Baptist, L. 2000. GENESIS-II: A language generation module for conversational systems. SM. Thesis, MIT, MA.

Baptist, I., Seneff, S. 2000. GENESIS-II: A versatile system for language generation in conversational system applications. *Proc. International Conference on Spoken Language Processing (ICSLP)*, pp. 271–274.

Barnard, P. J. 1987. Cognitive resources and the learning of human-computer dialogs. In *Interfacing Thought, Cognitive Aspects of Human-Computer Interaction*, J. M. Carroll (ed.), MIT Press, pp. 112–158.

Barnwell, T. P., Nayebi, K., Richardson, C. H. 1996. *Speech coding: A computer laboratory textbook*. Georgia Tech Digital Signal Processing Laboratory Series.

Bascle, B., Blake, A., Morris, J. 1998. Towards automated, real-time, facial animation. In Cipolla, R., Pentland, A. (eds). *Computer Vision for Human-Machine Interaction*. Cambridge University Press, pp. 123–133.

Basu, S., Oliver, N., Pentland, A. 1998. 3D modelling and tracking of human lip motion. *Proc. ICCV*, Bombay, India, pp. 337–343.

Bell, L., Boye, J., Gustafson, J., Wirén, M. 2000. Modality convergence in a multimodal dialogue system. *Proc. Cötalog 2000, 4th Workshop on the Semantics and Pragmatics of Dialogue*, pp. 29–34.

Bell, L., Gustafson, J. 1999. Interacting with an animated agent: An analysis of a Swedish database of spontaneous computer directed speech. *Proc. Eurospeech*, pp. 1143–1146.

Benoît, C., Mohamadi, T., Kandel, S. 1994. Effects of phonetic context on audio-visual intelligibility of French speech in noise. *Journal of Speech & Hearing Research*, 37, pp. 1195–1203.

Beringer, N., Kartal, U., Louka, K., Schiel, F., Türk, U. 2002b. PROMISE – A procedure for multimodal interactive system evaluation. *Proc. LREC Workshop on Multimodal Resources and Multimodal Systems Evaluation*, Las Palmas, Canary Islands, Spain.

Beringer, N., Katerina, L., Penide-López, V., Türk, U. 2002a. End-to-end evaluation of multimodal dialogue systems – can we transfer established methods?. *Proc. 3rd Language Resources and Evaluation Conference*, pp. 558–563.

Bertelson, P., Radeau, M. 1976. Ventriloquism, sensory interaction, and response bias: remarks on the paper by Choe, Welch, Gilford and Juola. *Perception and Psychophysics*, vol. 29, pp. 578–585.

Berthommier, F. 2003. Audiovisual speech enhancement based on the association between speech envelope and video features. *Proc. Eurospeech*, pp. 1045–1048.

Beskow, J. 1995. Rule-based visual speech synthesis. *Proc. Eurospeech*, pp. 299–302.

Beskow, J. 1997. Animation of talking agents. *Proc. ESCA Workshop on Audio-visual Speech Processing*, pp. 149–152.

Beskow, J. 1998. A tool for teaching and development of parametric speech synthesis. *Proc. Fonetik*, Stockholm, Sweden.

Beskow, J., McGlashan, S. 1997. Olga – A conversational agent with gestures. *Proc. IJCAI '97 Workshop on Animated Interface Agents – Making Them Intelligent*, Nagoya, Japan.

Bickmore, T., Cassell, J. 2001. A relational agent: A model and implementation of building user trust. *Proc. Computer-Human Interaction*, pp. 396–403.

Bickmore, T., Cassell, J. 2004. Social dialogue with embodied conversational agents. In J. van Kuppevelt, L. Dybkjaer and N. Bernsen (eds) *Natural Intelligent and Effective Interaction with Multimodal Dialogue Systems*. Kluwer Academic Press.

Billi, R., Castagneri, G., Danielli, M. 1997. Field trial evaluations of two different information inquiry systems. *Speech Communication*, vol. 23, no. 1–2, pp. 83–93.

Binot, J. L, Falzon, P., Perez, R., Peroche, B., Sheehy, N., Rouault, J., and Wilson, M. 1990. Architecture of a multimodal dialogue interface for knowledge-based systems. *Proc. ESPRIT 1990*, pp. 412–433, Kluwer.

Bolt, R. A. 1980. Put-That-There: Voice and gesture at the graphics interface. *Computer Graphics*, vol. 14(3), pp. 262–270.

Boros, M., Eckert, W., Gallwitz, F., Goerz, G., Hanrieder, G., Niemann, H. 1996. Towards understanding spontaneous speech: word accuracy vs. concept accuracy. *Proc. International Conference on Spoken Language Processing (ICSLP)* 96, pp. 1009–1012.

Borsley, R. 1996. *Modern Phrase Structure Grammar*. Blackwell Publishing.

Bour, L. 1997. *Dmi-search Scleral Coil*. Tech. Rep. H2-214, Dept. of Neurology, Clinical Neurophysiology, Academic Medical Centre, AZUA, Amsterdam, Netherlands.

Brennan, S. E. 1996. Lexical entrainment in spontaneous dialog. *Proc. International Symposium on Spoken Dialog*, pp. 41–44.

Brennan, S. E., Friedman, M. W., Pollard, C. J. 1987. A centering approach to pronouns. *Proc. 25th Annual Meeting of the ACL*, pp. 155–162.

Brøndsted, T. 1999. Reference problems in Chameleon. *Proc. Tutorial and Research Workshop on Interactive Dialogue in Multi-Modal Systems*, pp. 133–136.

Brøndsted, T., Dalsgaard, P., Larsen, L. B., Manthey, M., Mc Kevitt, P., Moeslund, T. B., Olesen, K. G. 2001. The IntelliMedia workbench – An environment for building multimodal systems. In H. Bunt and R.-J. Beun (eds). *Lecture Notes on Artificial Intelligence, 2155*, pp. 217–233. Springer-Verlag, Berlin.

Brooke, N. M. 1990. Visible speech signals: Investigating their analysis, synthesis and perception. In M. M. Taylor, F. Néel, and D. G. Bouwhuis (eds). *The Structure of Multimodal Dialogue*. Elsevier Science.

Brown, D. C., Kwasny, S. C., Chandrasekaran, B., Sondheimer, N. K. 1979. An experimental graphics system with natural-language input. *Computer and Graphics*, vol. 4, pp. 13–22.

Bühler, D., Minker, W., Häussler, J., Krüger, S. 2002. Flexible multimodal human-machine interaction in mobile environments. *Proc. ECAI workshop on Artificial Intelligence in Mobile Systems (AIMS)*, pp. 66–70.

Buisine, S., Abrilian, S., Rendu, C., Martin, J. C. 2002. Towards experimental specification and evaluation of lifelike multimodal behavior. *Proc. Workshop on embodied conversational agents – let's specify and evaluate them!*, pp. 42–48.

Burnard, L. 1995. *User's Reference Guide for the British National Corpus version 1.0*. Oxford University Computing Services, Oxford.

Caelen, J., Bruandet, M. F. 2001. Interaction multimodale pour la recherche d'information. In C. Kolski (ed.) *Environnements évolués et évaluation de l'IHM*, pp. 175–205. Hermès Science Publications, Paris.

Caldognetto, E. M., Poggi, I., Cosi, P., Cavicchio, F., Merola, G. 2004. Multimodal score: an ANVIL based annotation scheme for multimodal audio-video analysis. *Proc. International Conference on Language Resources and Evaluation*, pp. 29–33.

Carbonell, J. R. 1970. *Mixed-initiative man-computer dialogues*. BBN Report No. 1971, Bolt, Beranek and Newman.

Cathcart, N., Carletta, J., Klein, E. 2003. A shallow model of backchannel continuers in spoken dialogue. In *Proc. the 10th Conference of the European Chapter of the Association for Computational Linguistics (EACL10)*, p. 51–58.

Carletta, J. C. 1992. Risk-taking and recovery in task-oriented dialogue. Unpublished PhD, Edinburgh University, http://homepages.inf.ed.ac.uk/jeanc/carletta-thesis.ps

Carletta, J. C. 1996. Assessing agreement on classification tasks: the kappa statistic. *Computational Linguistics*, Vol. 22, no. (2), pp. 249–254.

Carlson, R., Granström, B. 1993. The Waxholm spoken dialogue system. *Proc. Eurospeech*, pp. 1867–1870.

Carpenter, R. 1992. *The Logic of Typed Feature Structures*. Cambridge University Press.

Cassell, J., Bickmore, T., Billinghurst, M., Campbell, L., Chang, K., Vilhálmsson, H., Yan, H. 1999. Embodiment in conversational interfaces: Rea. *Proc. Computer-Human Interaction*, pp. 520–527.

Cassell, J., Bickmore, T., Campbell, L., Hannes, V., Yan, H. 2000. Conversation as a system framework: Designing embodied conversational agents. In Cassell J., Sullivan J., Prevost S., Churchill E. (Eds) *Embodied Conversational Agents*. MIT Press.

Cassell, J., Thorisson, K. R. 1999. The power of a nod and a glance: envelope vs. emotional feedback in animated conversational agents. *Applied Artificial Intelligence*, Vol. 13, pp. 519–538.

Cassell, J., Vilhjlmsson, H., Bickmore, T. 2001. BEAT: The behavior expression animation toolkit. *Proc. SIGGRAPH*, pp. 477–486.

Cavazza, M. 2001. Representation and reasoning in a multimodal conversational character. *Proc. 2nd Workshop on Knowledge and Reasoning in Practical Dialogue Systems*, International Joint Conference on Artificial Intelligence.

Cerisara, C. 2003. The use of confidence values in vector-based call routing. *Proc. Eurospeech*, pp. 633–636.

Cerrato, L., Ekeklint, S. 2002. Different ways of ending human-machine dialogues. *Proc. Embodied conversational agents – Let's specify and evaluate them!*

Cheyer, A., Julia, L. 1999. Designing, developing and evaluating multimodal applications. *Proc. Computer-Human Interaction*.

Chu, C. C., Dani, T. H., Gadh, R. 1997. Multisensory interface for a virtual reality based computer aided design system, *Computer-Aided Design*, vol. 29, no. (10), pp. 709–725.

Chu-Carroll, J., Carpenter, B. 1999. Vector-based natural language call routing. *Computational Linguistics*, vol. 25, no. (3), pp. 361–388.

Claasen, W. 2000. *Using recall measurements and subjective ratings to assess the usability of railroad travel plans presented by telephone*. Technical report no. 123. NWO Priority Program on Language and Speech Technology.

Clark, H. H., Brennan, S. E. 1991. Grounding in communication. In L. B. Resnick, J. Levine, S. D. Teasley (eds.), *Perspective on Socially Shared Cognition*, pp. 127–149. American Psychological Association Press.

Clark, H. H., Schaefer, E. F. 1987. Collaborating on contributions to conversations. *Language and Cognitive Process*, vol. 2, pp. 1–23.

Cohen, P. R., Johnston, M., McGee, D. R., Oviatt, S. L., Pittman, J., Smith, I., Clow, J. 1997. QuickSet: Multimodal interaction for distributed applications. *Proc. 5th International Multimedia Conference*, ACM Press, pp. 31–40.

Cohen, W. 1996. Learning trees and rules with set-valued features. *Proc. 13th National Conference on Artificial Intelligence (AAAI)*, pp. 709–716.

Cole, R. A. 1999. Tools for research and education in speech science. *Proc. International Conference on Phonetic Sciences*, San Francisco.

Corradini, A., Mehta, M., Bernsen, N. O., Martin, J.-C., Abrilian, S. 2003. Multimodal input fusion in human-computer interaction. *Proc. NATO-ASI Conference on Data Fusion for Situation Monitoring, Incident Detection, Alert and Response Management*, Yerevan, Armenia.

Costantini, E., Pianesi, F., Cosi, P. 2004. Evaluation of synthetic faces: Human recognition of emotional facial displays. *Proc. Affective Dialogue Systems: Tutorial and Research Workshop*, pp. 276–287.

Crowley, J. L., Bedrune, J. M. 1994. Integration and control of reactive visual processes. *Proc. Third European Conf. Computer Vision*, pp. 47–58.

Crowley, J. L., Berard, F. 1997. Multi-modal tracking of faces for video communication. *Proc. IEEE Conf. Computer Vision and Pattern Recognition*, pp. 640–645.

Dahlbäck, N., Jönsson, A., Ahrenberg, L. 1993. Wizard of Oz studies – why and how. *Proc. Int. Workshop on Intelligent User Interfaces*, pp. 193–200.

Dale, R., Moisl, H., Somers, H. (eds) 2000. *Handbook of Natural Language Processing*. Dekker Publishers.

Daubias, P., Deléglise, P. 2002. Lip-reading based on a fully automatic statistical model. *Proc. International Conference on Spoken Language Processing (ICSLP)*, pp. 209–212.

De Carolis, N., Carofiglio, V., Pelachaud, C. 2002a. From discourse plans to believable behaviour generation. *Proc. International Natural Language Generation Conference*, pp. 65–72.

De Carolis, B., Carofiglio, V., Vilvi, M., Pelachaud, C. 2002b. APML, a mark-up language for believable behavior generation. *Proc. Workshop on embodied conversational agents – let's specify and evaluate them!*

Deco, G., Obradovic, D. 1996. *An Information Theoretic Approach to Neural Computing*. Springer-Verlag.

D'Haro, L. F., Córdoba, R. de, San-Segundo, R., Montero, J. M., Macías-Guarasa, J., Pardo, J. M. 2004. Strategies to reduce design time in multimodal/multilingual dialog applications. *Proc. International Conference of Spoken Language Processing (ICSLP)*, pp. IV-3057–3060.

Dehn, D., Van Mulken, S. 2000. The impact of animated interface agents: A review of empirical research. *Proc. International Journal of Human-Computer Studies*, vol. 52, pp. 1–22.

Denecke, M., Yang, J. 2000. Partial information in multimodal dialogue. *Proc. Third International Conference on Advances in Multimodal Interfaces*, pp. 624–633.

Duchnowski, P., Hunke, M., Büsching, D., Meier, U., Waibel, A. 1995. Toward movement-invariant automatic lip-reading and speech recognition. *Proc. Int. Conf. on Acoustics, Speech and Signal Processing*, pp. 109–111.

Duchowski, A. T. 2003. *Eye Tracking Methodology: Theory and Practice*. Springer Verlag.

Dudney, B., Lehr, J. 2003. *Jakarta Pitfalls: Time-Saving Solutions for Struts, Ant, Junit, and Cactus*, John Wiley & Sons.

Duneau, L., Dorizzi, B. 1996. On-line cursive script recognition: A user-adaptive system for word recognition. *Pattern Recognition*, vol. 29, no. (12), pp. 1981–1994.

Dutoit, T. 1996. *An Introduction to Text-to-Speech Synthesis*. Kluwer Academic Publishers.

Dybkjær, L., Bernsen, N. O. 2002. Data resources and annotation schemes for natural interactivity: Purposes and needs. *Proc. Workshop on Multimodal Resources and Multimodal Systems Evaluation*, LREC, pp. 1–7.

Dybkjær, L., Bernsen, N. O. 2004. Recommendations for natural interactivity and multimodal annotation schemes. *Proc. International Conference on Language Resources and Evaluation, Workshop on Multimodal Corpora*, Lisbon, Portugal, pp. 32–40.

Ekman, P. 1992. An argument for basic emotions. *Cognition and Emotion*, vol. 6, pp. 169–200.

Ekman, P., Friesen, W. 1978. *Facial Action Coding System*. Consulting Psychologist Press.

Ekman, P., Rosenberg, E. L. 1997. *What the Face Reveals: Basic and Applied Studies of Spontaneous Expression Using the Facial Action Coding System (FACS)*. Oxford University Press.

El Jed, M., Pallamin, N., Dugdale, J., Pavard, B. 2004. Modelling character emotion in an interactive virtual environment. *Proc. AISB symposium: Motion, Emotion and Cognition*, Leeds.

Elting, C., Zwickel, J., Malaka, R. 2002. Device-dependent modality selection for user interfaces. *Proc. 6th International Conference on Intelligent User Interfaces*. San Francisco.

Essa, I. A., Pentland, A. P. 1995. Facial expression recognition using a dynamic model and motion energy. *Proc. International Conference on Computer Vision*, pp. 360–367.

Fabri, M., Moore, D. J., Hobbs, D. J. 2002. Expressive agents: Non-verbal communication in collaborative virtual environments. *Proc. Autonomous Agents and Multi-Agent Systems (Embodied Conversational Agents)*, July 2002, Bologne, Italy.

Forbes-Riley, K., Litman, D. J. 2004. Predicting emotion in spoken dialogue from multiple knowledge sources. *Proc. Human Language Technology Conference of the North American Chapter of the ACL*, HLT-NAACL, pp. 201–208.

Fujie, S., Fukushima, K., Kobayashi, T. 2004. A conversation robot with back-channel feedback function based on linguistic and nonlinguistic information. *Proc. International Conference on Autonomous Robots and Agents*, ICARA2004, pp. 379–384.

Funakoshi, T., Tokunaga, T. 2003. Evaluation of a robust parser for spoken Japanese. *Proc. DiSS'03, Disfluency in Spontaneous Speech Workshop*. pp. 53–56.

Furnham, A. 1990. Language and personality. In H. Giles and W. P. Robinson (eds). *Handbook of Language and Social Psychology*. pp. 73–95. John Wiley and Sons.

Gao, Y., Gu, L., Kuo, H. K. 2005. Portability challenges in developing interactive dialogue systems, *Proc. ICASSP 2005*, SS-13.6.

Gaver, W. W. 1994. Using a creating auditory icons. In G. Kramer (ed.) *Auditory Display*, pp. 417–446, Santa Fe Institute, Addison-Wesley.

Gong, S., McKenna, S. J., Psarrou, A. 2000. *Dynamic Vision: From Images to Face Recognition*. Imperial College Press & World Scientific Publishing.

Gibbon, D., Mertins, I., Moore, R. K. 2000. *Handbook of multimodal and spoken dialogue systems: Resources, Terminology and Product Evaluation*. Kluwer Academic Publishers.

Gips, J., Olivieri, P., Tecce, J. 1993. Direct control of the computer through electrodes place around the eyes. *Proc. 5th International Conference on Human-Computer Interaction*, pp. 630–635.

Girin, L., Schwartz, J. L., Feng, G. 2001. Audio-visual enhancement in of speech in noise. *JASA*, vol. 6, no. (109), pp. 1224–1227.

Glass, J., Flammia, G., Goodine, D., Phillips, M., Polifroni, J., Sakai, S., Seneff, S., Zue, V. 1995. Multilingual spoken-language understanding in the MIT Voyager system. *Speech Communication*, vol. 17, no. (1–2), pp. 1–18.

Glass, J., Polifroni, J., Seneff, S., Zue, V. 2000. Data collection and performance evaluation of spoken dialogue systems: The MIT experience. *Proc. International Conference on Spoken Language Processing (ICSLP)*, pp. 1–4.

Goto, J., Kim, Y.-B., Miyazaki, M., Komine, K., Uratani, N. 2004. A spoken dialogue interface for TV operations based on data collected by using WOZ method. *IEICE Transactions on Information and Systems*, vol. E87-D, no. 6, pp. 1397–1404.

Gould, J. D., Conti, J. Hovanyecz, T. 1982. Composing letters with a simulated listening typewriter. *Proc. Computer-Human Interaction*, pp. 367–370.

Granström B., House D., Lundeberg M. 1999. Prosodic cues in multimodal speech perception. *Proc. ICPhS*, pp. 655–658.

Granström B., House D., Swerts M. G., 2002. Multimodal feedback cues in human-machine interactions. *Proc. the Speech Prosody 2002 Conference*, pp. 347–350.

Grice, H. 1969. Utterer's meaning and intentions. *Philosophical Review*, vol. 68, no. (2), pp. 147–177.

Grishman R., Kittredge R. 1986. *Analyzing Language in Restricted Domain: Sublanguage Description and Processing*. Erlbaum Associates.

Grosz, J. B., Joshi, A. K. and Weinstein, S. 1995. Centering: A framework for modeling the local coherence of discourse. *Computational Linguistics*, vol. 21, pp. 203–225.

Grosz, J. B., Sidner, C. 1986. Attentions, intentions, and the structure of discourse. *Computational Linguistics*, vol. 12, pp. 175–204.

Guiard-Marigny, T., Tsingos, N., Adjoundani, A., Benoît, C., Gascuel, M. P. 1996. 3D models of the lips for realistic speech animation. *Proc. Second ESCA/IEEE Workshop on Speech Synthesis*, pp. 49–52.

Gustafson, J., Bell, L., Beskow, J., Boye, J., Carlson, R., Edlund, J., Granström, B., House, D., Wirén, M. 2000. AdApt – a multimodal conversational dialogue system in an apartment domain. *Proc. International Conference on Speech and Language Processing*, pp. 134–137.

Gustafson, J., Bell, L., Boye, J., Edlund, J., Wirén, M. 2002. Constraint manipulation and visualisation in a multi-modal dialogue system. *Proc. Tutorial and Research Workshop on Multimodal Dialogue in Mobile Environments*, Kloster Irsee, Germany.

Gustafson, J., Lindberg, N., Lundeberg, M. 1999. The August spoken dialogue system. *Proc. Eurospeech*, pp. 1151–1154.

Haag, A., Goronzy, S., Schaich, P., Williams, J. 2004. Emotion recognition using bio-sensors: first steps towards an automatic system. *Proc. Affective Dialogue Systems: Tutorial and Research Workshop*, pp. 36–48.

Hamerich, S. W., Córdoba, R. de, Schless , V., d'Haro, L.F., Kladis, B., Schubert, V., Kocsis, O., Igel, S., Pardo, J. M. 2004b. The GEMINI platform: semi-automatic generation of dialogue applications. *Proc. International Conference of Spoken Language Processing (ICSLP)*, pp. IV-2629–2632.

Hamerich, S. W., Schubert, V., Schless, V., Córdoba, R., Pardo, J. M., D'Haro, L. F., Kladis, B., Kocsis, O., Igel, S. 2004a. Semi-automatic generation of dialogue applications in the GEMINI project. *Proc. Workshop for Discourse and Dialogue (SigDial)*, pp. 31–34.

Hardy, H., Strzalkowski, T., Wu, M., Ursu, C., Webb, N., Biermann, A., Bryce, R. B., Mckenzie, A. 2004. Data-driven strategies for an automated dialogue system. *Proc. 42nd Annual Meeting of the ACL*, pp. 71–78.

Hasida, K., Den, Y. 1999. A synthetic evaluation of dialogue systems. In Yorick Wilks (ed.), *Machine Conversations*. Kluwer Academic Publishers, pp. 113–125.

Healey, J., Picard, R. 1998. Digital processing of affective signals. *Proc. Int. Conf. on Acoustics, Speech and Signal Processing*, Seattle, WA.

Heeman, P. A., Johnston, M., Denney, J., Kaiser, E. 1998. Beyond structured dialogues: factoring out grounding. *Proc. Internal Conference on Spoken Language Processing*, pp. 863–866.

Henton, C., Litwinowicz, P. 1994. Saying and seeing it with feeling: Techniques for synthesizing visible, emotional speech. *Proc. 2nd ESCA/IEEE Workshop on Speech Synthesis*, pp. 73–76.

Heracleous, P., Shimizu, T. 2003. An efficient keyword spotting technique using a complementary language for filler models training. *Proc. Eurospeech*, pp. 921–924.

Hershey, J., Movellan J. 1999. Using audio-visual synchrony to locate sounds. In S. A. Solla, T. K. Leen, K. R. Miller (eds) *Advances in Neural Information Processing Systems*, vol. 12, pp. 813–819.

Heylen, D., Vanes, I., Nijholt, A., Van Dijk, B. 2002. Experimenting with the gaze of a conversational agent. *Proc. Int. CLASS Workshop on Natural, Intelligent and Effective Interaction in Multimodal Dialogue Systems, Denmark*, pp. 93–100.

Hill, D., Pearce, A., Wyvill, B. 1998. Animating speech: an automated approach using speech synthesis by rules. *The Visual Computer*, vol. 3, pp. 277–289.

Hill, W., Wroblewski, D., McCandless, T., Cohen, R. 1992. Architectural qualities and principles for multimodal and multimedia interfaces. In *Multimedia Interface Design,* ACM Press, pp. 311–318.

Hirschman, L. 1994. Session 3: Human Language Evaluation. *Proc. Human Language Technology*. H94-1017.

Holzapfel, H., Fuegen, C., Denecke, M., Waibel, A. 2002. Integrating emotional cues into a framework for dialogue management. *Proc. Int. Conference on Multimodal Interfaces*, pp. 141–148.

Holzapfel, H., Nickel, K., Stiefelhagen, R. 2004. Implementation and evaluation of a constrained-based multimodal fusion system for speech and 3D pointing gestures. *Proc. International Conference on Multimodal Interfaces*, pp. 13–15.

Hone, K. S., Baber, C. 1995. Using a simulation method to predict the transaction time effects of applying alternative levels of constraint to utterances within speech interactive dialogues. *Proc. ESCA Workshop on Spoken Dialogue Systems*, pp. 209–212.

House, D., Beskow, J., Granström, B. 2001. Timing and interactions of visual cues for prominence in audiovisual speech perception. *Proc. Eurospeech*, pp. 387–390.

Hovy, E. H. 1993. Automated discourse generation using discourse relations. *Artificial Intelligence*, 63, Special Issue on Natural Language Processing, pp. 341–385.

Hsu, R.-L., Abdel-Mottaleb, M., Jain, A. K. 2002. Face detection in color images. *IEEE Trans. on Pattern Analysis and Machine Intelligence*, pp. 696–706.

Hu, C., Meng, M. Q., Liu, P. X., Wang, X. 2003. Visual gesture recognition for human-machine interface of robot teleoperation. *Proc. IEEE/RSJ International Conference on Intelligent Robots and Systems*, Las Vegas, Nevada, pp. 1560–1565.

Huang, X., Acero, A., Hon, H. 2001. *Spoken Language Processing: A Guide to Theory, Algorithm and System Development*. Prentice Hall.

Hutchins, W., Somers, H. 1992. *An Introduction to Machine Translation*. Academic Press.

Isbister, K., Nass, C. 1998. Personality in conversational characters: building better digital interaction partners using knowledge about human personality preferences and perceptions. *Proc. Workshop on Embodied Conversational Characters*, pp. 103–111.

JDRI (Japanese Discourse Research Initiative) 2000. Japanese dialogue corpus of multi-level annoatation. *Proc. of the 1st SIGDIAL Workshop on Discourse and Dialogue*.

Johnston, M., Bangalore, S. 2002. Finite-state multimodal parsing and understanding. *Proc. International Conference on Computational Linguistics*, Saarbrücken, Germany, pp. 369–375.

Johnston, M., Bangalore, S., Vasireddy, G., Stent, A., Ehlen, P., Walker, M., Whittaker, S., Maloor, P. 2002. MATCH: An architecture for multimodal dialogue systems. *Proc. 40th Annual Meeting of the ACL*, Philadelphia, pp. 376–383.

Johnston, M., Cohen, P. R., McGee, D., Oviatt, S. L., Pittman, J. A., Smith, I. 1997. Unification-based multimodal integration. *Proc. 35th Meeting of the ACL*, ACL Press, pp. 281–287.

Kaiser, E. C., Cohen, P. R. 2002. Implementation testing of a hybrid symbolic/statistical multimodal architecture. *Proc. International Conference on Spoken Language Processing (ICSLP)*, pp. 173–176.

Kamio, H., Kohrida, M., Matsu'ura, H., Tamura, M., Nitta, T. 1994. A UI design support tool for multimodal spoken dialogue system. *Proc. International Conference on Spoken Language Processing*, pp. 1283–1286.

Karlsson, I., Faulkner, A., Salvi, G. 2003. SYNFACE – a talking face telephone. *Proc. Eurospeech*, pp. 1297–1300.

Kellner, A., Rueber, B., Seide, F., Tran, B.-H. 1997. PADIS – An automatic telephone switchboard and directory information system. *Speech Communication*, vol. 23, pp. 95–111.

Kettebekov, S., Sharma, R. 2000. Understanding gestures in multimodal human computer interaction. *Proc. Int. Journal on Artificial Intelligence Tools*, pp. 205–224.

Kettebekov, S., Sharma, R. 2001. Toward natural gesture/speech control of a large display. In L. Nigay (Ed.), *Engineering for Human Computer Interaction*, vol. 2254, *Lecture Notes in Computer Science*, Springer Verlag, pp. 133–146.

Kettebekov, S., Yeasin, M., Krahnstoever, N., Sharma, R. 2002. Prosody based co-analysis of deictic gestures and speech in weather narration broadcast. *Proc. Multimodal Resources and Multimodal Systems Evaluation Workshop*, pp. 57–62.

Kieras, D., Polson, P. G. 1985. An approach to the formal analysis of user complexity. *Proc. International Journal of Man-Machine Studies*, vol. 22, pp. 365–394.

Kiesler, S., Sproull, L. 1997. 'Social' human-computer interaction. In B. Friedman (ed.). *Human Values and the Design of Computer Technology*, CSLI Publications, pp. 191–199.

Kipp, M. 2001. Anvil – A generic annotation tool for multimodal dialogue. *Proc. Eurospeech*, pp. 1367–1370.

Klein, J., Moon, Y., Picard, R. 2002. This computer responds to user frustration: Theory, design and results. *Interacting with Computers*, vol. 14, pp. 119–140.

Koda, T., Maes, P. 1996. Agents with faces: The effect of personification. *Proc. 5th IEEE Int. Workshop on Robot and Human Communication*, pp. 189–194.

Komatani, K., Tanaka, K., Kashima, H., Kawahara, T. 2001. Domain-independent spoken dialogue platform using key-phrase spotting based on combined language model. *Proc. Eurospeech*, pp. 1319–1322.

Kramer, G. 1994. An introduction to auditory display. In G. Kramer (ed.). *Auditory Display*, Santa Fe Institute, Addison-Wesley. pp. 1–77.

Kranstedt, A., Kopp, S., Wachsmuth, I. 2002. MURML: A multimodal utterance representation markup language for conversational agents. *Proc. Workshop on Embodied Conversational Agents, Autonomous Agents and Multi-Agent Systems*, Bologne, Italy.

Krum, D. M., Omoteso, O., Ribarski, W., Starner, T., Hodges, L. F. 2002. Evaluation of a multimodal interface for 3D terrain visualization. *IEEE Visualization*, pp. 411–418.

Krumm, J., Harris, S., Meyers, B., Brummit, B., Hale, M. Shafer, S. 2000. Multi-camera multi-person tracking for easyliving. *Proc. 3rd IEEE Workshop on Visual Surveillance*, pp. 3–10.

Kshirsagar, S., Magnenat-Thalman, N. 2002. A multilayer personality model. *Proc. 2nd International Symposium on Smart Graphics*, pp. 107–115.

Lakoff, R. T. 1973. The logic of politeness; or minding your p's and q's. *Proc. 9th Regional Meeting of the Chicago Linguistic Society*, pp. 292–305.

Lamar, M. V., Bhuiyan, M. S., Iwata, A. 1999. Hand gesture recognition analysis and an improved CombNET-II. *Proc. IEEE International Conference on Systems, Man and Cybernetics*, vol. 4, pp. 57–62.

Lang, P. J. 1995. The emotion probe: Studies of motivation and attention. *American Psychologist*, vol. 50, no. (5), pp. 372–385.

Lang, P. J., Bradley, M. M., Cuthbert, B. N. 1990. Emotion, attention, and startle reflex. *Psychological Review*, vol. 97, no. (3), pp. 377–395.

Larson, J.A. 2002. *Voicexml: Introduction to Developing Speech Applications*, Prentice Hall.

Lascarides, A., Briscoe, T., Asher, N., Copestake, A. 1996. Order independent and persistent typed default unification. *Linguistics and Philosophy*, vol. 19, no. (1), pp. 1–89.

Lavagetto, F., Pockaj, R. 1999. The facial animation engine: Towards a high-level interface for design of MPEG-4 compliant animated faces. *IEEE Transactions on Circuits and Systems for Video Technology*, vol. 9, no. (2), pp. 277–289.

LeCun, Y., Jackel, L., Bottou, L., Brunot, A., Cortes, C., Denker, J., Drucker, H., Guyon, I., Miiller, U., Säckinger, E., Simard, P., Vapnik, V. 1995. Comparison of learning algorithms for handwritten digit recognition. *Proc. International Conference on Artificial Neural Networks*, pp. 53–60.

Lee, A., Kawahara, T., Shikano, K. 2001. Julius - an open source real-time large vocabulary recognition engine. *Proc. European Conf. on Speech Communication and Technology*, pp. 1691–1694.

Lee, C., Xu, Y. 1996. Online, interactive learning of gestures for human/robot interfaces. *Proc. IEEE International Conference on Robotics and Automation*, pp. 2982–2987.

Lee, C.-H., Carpenter, B., Chou, W., Chu-Carroll, J., Reichl, W., Saad, A., Zhou, Q. 2000. On natural language call routing. *Speech Communication*, vol. 31, pp. 309–320.

Lee, C. M., Narayanan, S., Pieraccini, R. 2001. Recognition of negative emotions from the speech signal. *Proc. IEEE ASRU*, pp. 240–243.

LeGoff. B., Guiard-Marigny, T., Benoît, C. 1996. Analysis-synthesis and intelligibility of a talking face. In J. van Santen, R. Sproat, J. Olive, and J. Hirschberg (eds) *Progress in Speech Synthesis*, Springer Verlag.

Lemon, O., Bracy, A., Gruenstein, A., Peters, S. 2001. The WITAS Multi-Modal Dialogue System I. *Proc. Eurospeech*, pp. 1559–1562.

Levin, E., Pieraccini, R., Eckert, K. 2000. A stochastic model of human-machine interaction for learning dialogue strategies. *IEEE Transactions on Speech and Audio Processing*, vol. 8, no. (1), pp. 11–23.

Lewis, C., Polson, P., Wharton, C., Rieman, J. 1990. Testing a walkthrough methodology for theory-based design of walk-up-and-use interfaces. *Proc. Computer-Human Interaction*, pp. 235–241.

Life, A., Salter, I., Temem, J. M., Bernard, F., Rosset, S., Bennacef, S., Lamel, L. 1996. Data collection of the MASK kiosk: WOz vs Prototype system. *Proc. International Conference on Spoken Language Processing (ICSLP)*, pp. 1672–1675.

Litman, D. J., Pan, S. 2000. Predicting and adapting to poor speech recognition in a spoken dialogue system. *Proc. 17th National Conference on Artificial Intelligence (AAAI)*, pp. 722–728.

Liu, X., Zhao, Y., Pi, X., Liang, L., Nefian, A. V. 2002. Audio-visual continuous speech recognition using a coupled hidden Markov model. *Proc. International Conference on Spoken Language Processing (ICSLP)*, pp. 213–216.

Liu, Y. 2003. Automatic phone set extension with confidence measure for spontaneous speech. *Proc. Eurospeech*, pp. 2741–2744.

López-Cózar, R., García, P., Díaz, J., Rubio, A. J. 1997. A voice-activated dialogue system for fast-food applications. *Proc. Eurospeech*, pp. 1783–1786.

López-Cózar, R., De la Torre, A., Segura, J. C., Rubio, A. J., Sánchez, V. 2003. Assessment of dialogue systems by means of a new simulation technique. *Speech Communication*, vol. 40, no. (3), pp. 387–407.

Löwgren, J., Nordqvist, T. 1990. A Knowledge-based tool for user interface evaluation and its integration in a UIMS. *Human-Computer Interaction – INTERACT '90*, pp. 395–400.

Ma, J., Yan, J., Cole, R. 2002. CU Animate: Tools for enabling conversations with animated agents. *Proc. International Conference on Spoken Language Processing (ICSLP)*, pp. 197–200.

Macherey, W., Ney, H. 2003. A comparative study on maximum entropy and discriminative training for acoustic modeling in automatic speech recognition, *Proc. Eurospeech*, pp. 793–796.

Maloor, P., Chai, J. 2000. Dynamic user level and utility measurement for adaptive dialog in help-desk system. *Proc. 1st SIGdial Workshop on Discourse and Dialogue*, pp. 94–101.

Mariani, J. 1993. Overview of the COCOSDA initiative. *Proc. Workshop of the International Coordinating Committee on Speech Databases and Speech I/O Assessment*.

Martell, C., Osborn, C., Friedman, J., Howard, P. 2002. The FORM gesture annotation system. *Proc. International Conference on Language Resources and Evaluation (LREC): Multimodal Resources and Multimodal Systems Evaluation Workshop*, Las Palmas, Canary Islands, Spain.

Martin, J., Veldman, R., Beéroule, D. 1995. Towards adequate representation technologies for multimodal interfaces. *Proc. International Conference on Cooperative Multimodal Communication*, pp. 207–223.

Martin, J.-C., Grimard, S., Alexandri, K. 2001. On the annotation of the multimodal behavior and computation of cooperation between modalities. *Proc. Workshop on Representing, Annotating and Evaluating Non-Verbal and Verbal Communicative Acts to Achieve Contextual Embodied Agents*, 5th International Conference on Autonomous Agents, pp. 1–7.

Martin, J.-C., Julia, L., Cheyer, A. 1998. A theoretical framework for multimodal user studies. *Proc. Second International Conference on Cooperative Multimodal Communication*, pp. 104–110.

Martin, J.-C., Kipp, M. 2002. Annotating and measuring multimodal behaviour – Tycoon metrics in the Anvil tool. *Proc. 3rd Language Resources and Evaluation Conference*, Las Palmas, Canary Islands, Spain.

Massaro, D. 2002. The psychology and technology of talking heads in human-machine interaction. *Proc. Int. CLASS Workshop on Natural, Intelligent and Effective Interaction in Multimodal Dialogue Systems*, pp. 106–119.

McAllister, D., Rodman, R., Bitzer, D., Freeman, A. 1997. Lip synchronization of speech. *Proc. AVSP '97*, Rhodes, Greece, pp. 133–136.

McBreen, H., Jack, M. 2001. Evaluating humanoid synthetic agents in e-retail applications. *IEEE SMC Transactions*, Special Issue on Socially Intelligent Agents, vol. 31, no. (5), pp. 394–405.

McCrae, R. R., John, O. P. 1992. An introduction to the five-factor model and its applications. Special Issue: The five-factor model: Issues and applications. *Journal of Personality*, vol. 60, pp. 175–215.

McGee, D. R., Cohen, P. R., Oviatt, S. 1998. Confirmation in multimodal systems. *Proc. International Joint Conference of the Association for Computational Linguistics and the International Committee on Computational Linguistics*, Montreal, Quebec, Canada, pp. 823–829.

McGurk, H., McDonald, J. 1976. Hearing lips and seeing voices. *Nature*, vol. 264, pp. 746–748.

McKevitt, P. 2003. Multimodal semantic representation. *Proc. SIGSEM Working Group on the Representation of Multimodal Semantic Information*, First Working Meeting, Fifth International Workshop on Computational Semantics (IWCS-5), Harry Bunt, Kiyong Lee, Laurent Romary, and Emiel Krahmer (Eds). Tilburg University, Tilburg, The Netherlands.

McNeill, D. 1992. *Hand and Mind: What Gestures Reveal about Thought*. The University of Chicago Press.

McTear, M. F. 1999. Software to support research and development of spoken dialogue systems. *Proc. Eurospeech*, pp. 339–342.

McTear, M. F. 2004. *Spoken Dialogue Technology: Toward the Conversational User Interface*. Springer.

Meier, U., Stiefelhagen, R., Yang, J., Waibel, A. 2000. Towards unrestricted lip reading. *International Journal on Pattern Recognition and Artificial Intelligence*, vol. 14, no. (5), pp. 571–585.

Mellor, B., Baber, C. 1997. Modelling of speech-based user interfaces. *Proc. Eurospeech*, pp. 2263–2266.

Meng, H., Ching, P. C., Wong, Y. F., Cheong Chat, C. 2002. ISIS: a multi-modal, trilingual, distributed spoken dialog system developed with CORBA, java, XML and KQML. *Proc. International Conference on Spoken Language Processing (ICSLP)*, pp. 2561–2564.

Meteer, M. et al. 1995. Disfluency annotation stylebook for the switchboard corpus.http://www.ldc.upenn.edu/Catalog/docs/treebank3/DFLGUIDE.PDF.

Ming Cheng, Y., Liu, C., Wei, Y. J., Melnar, L. and Ma, C. 2003. An approach to multilingual acoustic modeling for portable devices. *Proc. Eurospeech*, pp. 3121–3124.

Montero, J. A., Sucar, L. E. 2004. Feature selection for visual gesture recognition using Hidden Markov models. *Proc. Fifth International Conference in Computer Science*, pp. 196–203.

Moran, D. B., Cheyer, A. J., Julia, L. E., Martin, D. L., Park, S. 1997. Multimodal user interfaces in the Open Agent Architecture. *Proc. International Conference on Intelligent User Interfaces*, pp. 61–68.

Moreno, R., Mayer, R. E., Spires, H. A., Lester, J. C. 2001. The case for social agency in computer-based teaching: do students learn more deeply when they interact with animated pedagogical agents? *Cognition and Instruction*, vol. 19, pp. 177–213.

Mori, S., Nishimura, M., Itoh, N. 2003. Language model adaptation using word clustering. *Proc. Eurospeech*, pp. 425–428.

Murao, H., Kawaguchi, N., Matsubara, S., Yamaguchi, Y., Inagaki, Y. 2003. Example-based spoken dialogue system using WOz system log. *Proc. 4th ACL SigDial Workshop on Discourse and Dialogue*, pp. 140–148.

Nahas, M., Huitric, H., Saintourens, M. 1998. Animation of a B-Spline figure. *The Visual Computer*, 3(5), pp. 272–276.

Nakano N., Minami Y., Seneff S., Hazen T. J., Cyphers D. S., Glass J., Polifroni J., Zue V. 2001. Mokusei: A telephone-based Japanese conversational system in the weather domain. *Proc. Eurospeech*, pp. 1331–1334.

Neal, J. G., Shapiro, S. C. 1991. Intelligent multi-media interface technology. Sullivan J. V. and Tyler S. W. (Eds) *Intelligent User Interfaces*, ACM Press, pp. 11–43.

Nigay, L., Coutaz, J. 1995. A generic platform for addressing the multimodal challenge. *Proc. the SIGCHI Conference on Human Factors in Computing Systems*, ACM, pp. 98–105.

Nigay, L., Coutaz, J., Salber, D. 1993. MATIS: A multi-modal airline travel information system. The AMODEUS project – ESPRIT BRA 7040, Report SM/WP 10.

Niimi, Y., Nishimoto, T., Kobayashi, Y. 1997. Analysis of interactive strategy to recover from misrecognition of utterances including multiple information items. *Proc. Eurospeech 97*, pp. 2251–2254.

Niimi, Y., Oku, T., Nishimoto, T., Araki, M. 2000. A task-independent dialogue controller based on the extended frame-driven method. *Proc. International Conference on Spoken Language Processing (ICSLP)*, pp. 114–117.

Nock, H. J., Iyengar, G., Neti, C. 2002. Assessing face and speech consistency for monologue detection in video. *Proc. ACM Multimedia*, Juan-les-Pins, France, pp. 303–306.

Ohno, T., Mukawa, N., Kawato, S. 2003. Just blink your eyes: A head-free gaze tracking system. *Proc. Computer-Human Interaction*, pp. 950–951.

Omukai, J., Araki, M. 2004. Construction of multilingual voice portal system, Tech. report of IPSJ, 2004-SLP-54-55 (in Japanese), pp. 325–330.

Ortony, A., Clore, G. L., Collins, A. 1988. *The Cognitive Structure of Emotions*. Cambridge University Press.

Oviatt, S. L. 1996. Multimodal interfaces for dynamic interactive maps. *Proc. Conference Human Factors in Computing Systems*, pp. 95–102.

Oviatt, S. 2000. Taming recognition errors with a multimodal interface. *Communications of the ACM*, 43(9), pp. 45–51.

Oviatt, S., Olsen, E. 1994. Integration themes in multimodal human-computer interaction. *Proc. International Conference on Spoken Language Processing (ICSLP)*, pp. 551–554.

Parke, F. I. 1982. Parametrized models for facial animation. *IEEE Computer Graphics*, 2(9), pp. 61–68.

Pastoor, S., Liu, J., Renault, S. 1999. An experimental multimedia system allowing 3-d visualization and eye-controlled interaction without user-worn devices. *IEEE Transactions on Multimedia*, vol. 1, no. 1, pp. 41–52.

Patel, M., Willis, P. J. 1991. FACES: The facial animation, construction and editing system. *Proc. Eurographics*, pp. 33–45.

Pavlovic, V., Sharma, R., Huang, T. 1997. Visual interpretation of hand gestures for human-computer interaction: A review. *IEEE Transactions on Pattern Analysis and Machine Intelligence*, vol. 19, no. 7, pp. 677–695.

Pelachaud, C., Prevost, S. 1995. Coordinating vocal and visual parameters for 3D virtual agents. *Proc. 2nd Eurographics Workshop on Virtual Environments*, pp. 99–106.

Pelachaud, C., Viaud, M., Yahia, H. 1993. Rule-structured facial animation system. *Proc. IJCAI*, pp. 1610–1615.

Pfleger, N., Alexandersson, J., Becker, T. 2002. Scoring functions for overlay and their application in discourse processing. *Proc. KOVENS*, pp. 139–146, Saarbrücken, Germany.

Pfleger, N., Alexandersson, J., Becker, T. 2003. A robust and generic discourse model for multimodal dialogue. *Proc. 3rd IJCAI workshop on Knowledge and Reasoning in Practical Dialogue Systems*.

Picard, R. W. 2000. *Affective Computing*. MIT Press.

Platt, S., Badler, N. 1981. Animating facial expressions. *Computer Graphics*, vol. 15, no. (3), pp. 245–252.

Plumbley, M. 1991. *On information theory and unsupervised neural networks*. Technical report CUED/F-INFENG/TR. 78, Cambridge University Engineering Department, UK.

Polifroni, J., Seneff, S., Glass, J., Hazen, T. J. 1998. Evaluating methodology for a telephone-based conversational system. *Proc. First Language Resources and Evaluation Conference*, pp. 43–49.

Potjer, J., Russel, A., Boves, L., Den Ons, E. 1996. Subjective and objective evaluation of two types of dialogues in a call assistance service. *Proc. IVTTA*, pp. 121–124.

Prendinger, H., Mayer, S., Mori, J., Ishizuka, M. 2003. Persona effect revisited. Using bio-signals to measure and reflect the impact of character-based interfaces. *Proc. 4th International Working Conference on Intelligent Virtual Agents*, pp. 283–291.

Prendinger, H., Dohi, H., Wang, H., Mayer, S., Ishizuka, M. 2004. Empathic embodied interfaces: Addressing user's affective state. *Proc. Affective Dialogue Systems: Tutorial and Research Workshop*, pp. 53–64.

Prodanov, P., Drygajlo, A. 2003. Bayesian networks for spoken dialogue management in multimodal systems of tour-guide robot. *Proc. Eurospeech*, pp. 1057–1060.

Puckette, M. 1995. Formant-based audio synthesis using nonlinear distortion. *Audio Engineering Society*, Reprinted from JAES 43/1, pp. 40–47.

Rabiner, L. R., Juang, B. H. 1993. *Fundamentals of Speech Recognition*. Prentice-Hall.

Rabiner, L. R., Juang, B. H., Lee, C. H. 1996. An overview of automatic speech recognition. *Automatic Speech and Speaker Recognition: Advanced Topics*, Kluwer Academic Publisher, pp. 1–30.

Reeves, B., Nass, C. 1996. *The Media Equation*. Cambridge University Press.

Reithinger, N., Lauer, C., Romary, L. 2002. MIAMM – Multidimensional information access using multiple modalities. *Proc. International CLASS Workshop on Natural Intelligent and Effective Interaction in Multimodal Dialogue Systems*.

Revéret, L., Benoît, C. 1998. A new 3D lip model for analysis and synthesis of lip motion in speech production. *Proc. AVSP*, Terrigal, Australia, pp. 207–212.

Riccardi, G., Gorin, A. L., Ljolje, A., Riley, M. 1997. A spoken language system for automated call routing. *Proc. Int. Conf. on Acoustics, Speech and Signal Processing*, pp. 1143–1146.

Risberg, A., Lubker, J. L. 1978. *Prosody and Speechreading*. Quarterly Progress & Status Report 4, Speech Transmission Laboratory, KTH, Stockholm, Sweden.

Roe, D., Pereira, F., Sproat, R., Riley, M., Moreno, P., Macarron, A. 1991. Toward a spoken language translator for restricted-domain context-free languages. *Proc. Eurospeech*, pp. 1063–1066.

Rogozan, A., Deléglise, P. 1998. Adaptive fusion of acoustic and visual sources for automatic speech recognition. *Speech Communication*, vol. 26, no. (1–2), pp. 149–161.

Rowley, H. A., Baluja, S., Kanade, T. 1998. Neural network-based face detection. *IEEE Transactions on Pattern Analysis and Machine Intelligence*, vol. 20, no. (1), pp. 23–28.

Rüber, B. 1997. Obtaining confidence measures from sentence probabilities. *Proc. Eurospeech*, pp. 739–742.

Salber, D., Coutaz, J. 1993. Applying the Wizard of Oz technique to the study of multimodal systems. In *Human Computer Interaction*, 3rd International Conference EWHCI'93, L. Bass, J. Gornostaev, C. Unger (Eds) Springer-Verlag, *Lecture notes in Computer Science*, Vol. 753, 1993, 219–230.

Salber, D., Coutaz, J., Nigay, L. 1993. VoicePaint: A voice and mouse controlled drawing program. *The AMODEUS project* – ESPRIT BRA 7040, Report SM/WP 9.

Salisbury, M. W., Hendrickson, J. H., Lammers, T. L., Fu, C., Moody, S. A. 1990. Talk and draw: bundling speech and graphics. *IEEE Computer*, vol. 23, no. (8), pp. 59–65.

Samal, A., Iyengar, P. A. 1995. Human face detection using silhouettes. *International Journal on Pattern Recognition and Artificial Intelligence*, vol. 9, no. (6), pp. 845–867.

Sanders, G. A., Le, A. N., Garafolo, J. S. 2002. Effects of word error rate in the DARPA Communicator data during 2000 and 2001. *Proc. International Conference on Spoken Language Processing (ICSLP)*, pp. 277–280.

Sasou, A., Asano, F., Tanaka, K., Nakamura, S. 2003. Adaptation of acoustic model using the gain-adapted HMM decomposition method. *Proc. Eurospeech*, pp. 29–32.

Satta, G. 2000. Parsing techniques for lexicalized context-free grammars. *Proc. 6th IWPT*, Trento, Italy.

Scherer, K. R. 1979. Personality markers in speech. In K. R. Scherer & H. Giles (eds). *Social Markers in Speech*. Cambridge University Press, pp. 147–209.

Scherer, K., Ekman, P. 1982. *Handbook of Methods in Nonverbal Behavior Research*. Cambridge University Press.

Schiel, F., Steininger, S., Beringer, N., Türk, U., Rabold, S. 2002. Integration of multi-modal data and annotations into a simple extendable form: The extensions of the BAS partitur format. *Proc. LREC Workshop on Multimodal Resources and Multimodal Systems Evaluation*, Las Palmas, Canary Islands, Spain.

Schiel, F., Steininger, S., Türk, U. 2003. *The SmartKom multimodal corpus at BAS*. bmb+f report no.34.

Schmidt Feris, R., Emídio de Campos, T., Marcondes Cesar, R. 2000. Detection and tracking of facial features in video sequences. *Lecture Notes in Artificial Intelligence*, vol. 1793, pp. 197–206.

Schramm, H., Rueber, B., Kellner, A. 2000. Strategies for name recognition in automatic directory assistance systems. *Speech Communication*, 31, pp. 329–338.

Schweiger, R., Bayerl, P., Neumann, H. 2004. Neural architecture for temporal emotion classification. *Proc. Affective Dialogue Systems: Tutorial and Research Workshop*, pp. 49–52.

Searle, J. R. 1975. *Language, Mind and Knowledge*. University of Minnesota Press.

Seneff, S. 1989. TINA: A probabilistic syntactic parser for speech understanding systems. *Proc. ICASSP*, pp. 711–714.

Seneff, S. 1992a. TINA: A natural language system for spoken language applications. *Computational Linguistics*, vol. 18, no. (1), pp. 61–86.

Seneff, S. 1992b. Robust parsing for spoken language systems. *Proc. ICASSP*, pp. 189–192.

Seneff, S., Hurley, E., Lau, R., Pao, C., Schmid, P., Zue, V. 1998. Galaxy-II: A reference architecture for conversational system architecture. *Proc. International Conference on Spoken Language Processing (ICSLP)*, pp. 931–934.

Seneff, S., Meng, H., Zue, V. 1992. Language modelling for recognition and understanding using layered bigrams. *Proc. International Conference on Spoken Language Processing (ICSLP)*, pp. 317–320.

Seneff S., Polifroni J. 2000. Dialogue management in the Mercury flight reservation system. *Proc. ANLP-NAACL.* Satellite Workshop, pp. 1–6.

Serrah, J., Gong, S. 2000. Resolving visual uncertainty and occlusion through probabilistic reasoning. *Proc. British Machine Vision Conference*, vol. 1, Univ. of Bristol, September, pp. 252–261.

Seto, S., Kanazawa, H., Shinchi, H., Takebayashi, Y. 1994. Spontaneous speech dialogue system TOSBURG II and its evaluation. *Speech Communication*, no. 15, pp. 341–353.

Shamaie, A., Sutherland, A. 2003. Accurate recognition of large number of hand gestures. 2nd Iranian Conference on Machine Vision and Image Processing, K.N. Toosi University of Technology, 13–15 February, Tehran, Iran.

Sharma, R., Cai, J., Chakravarthy, S., Poddar, I., Sethi, Y. 2000. Exploiting speech/gesture co-occurrence for improving continuous gesture recognition in weather narration. *Proc. International Conference on Face and Gesture Recognition*.

Shih, S.-W., Wu, Y.-T., Liu, J. 2000. A calibration-free gaze tracking technique. *Proc. 15th Conference on Pattern Recognition*, pp. 4201–4204.

Shimoga, K. B. 1993. A survey of perceptual feedback issues in Dexterous telemanipulation: Part II. Finger Touch Feedback. In *Proc. the IEEE Virtual Reality Annual International Symposium*, pp. 271–279.

Sidner C. L., Forlines C. 2002. Subset language for conversing with collaborative interface agents. *Proc. International Conference on Spoken Language Processing (ICSLP)*, pp. 281–284.

Sibert, L. 2000. Evaluation of eye gaze interaction. Proc. *Human Computer Interaction*, pp. 281–288.

Singh, S., Litman, D., Kearns, M., Walker, M. 2002. Optimizing dialogue management with reinforcement learning: Experiments with the NJFun system. *Journal of Artificial Intelligence Research*, vol. 16, pp. 105–133.

Skantze, G. 2002. Coordination of referring expressions in multimodal human-computer dialogue. *Proc. International Conference on Spoken Language Processing (ICSLP)*, pp. 553–556.

Solon, A., McKevitt, P., Curran, K. 2004. Mobile multimodal dynamic output morphing tourist systems. *Proc. First Annual International Conference on Broadband Networks*, October 25–29, San Jose, CA, USA.

Souvignier, B., Kellner, A., Rueber, B., Schramm, H., Seide, F. 2000. The thoughtful elephant: Strategies for spoken dialog systems. *IEEE Transactions on Speech and Audio Processing*, vol. 8, no. 1, pp. 51–62.

Sparrel, C. J. 1993. Coverbal iconic gesture in human-computer interaction. Ms thesis, MIT.

Srihari, R., Baltus, C. 1993. Incorporating syntactic constraints in recognizing handwritten sentences. *Proc. Int. Joint Conference on Artificial Intelligence*, pp. 1262–1267.

Steedman, M. 1991. Structure and intonation. *Language*, vol 67, pp. 262–296.

Steininger, S., Rabold, S., Dioubina, O., Schiel F. 2002. Development of the user-state conventions for the multimodal corpus in SmartKom. *Proc. LREC Workshop on Multimodal Resources and Multimodal Systems Evaluation*, Las Palmas, Canary Islands, Spain, pp 33–37.

Steininger, S., Schiel, F., Louka, K. 2001. Gestures during overlapping speech in multimodal human-machine dialogues. *Proc. International workshop on information Presentation and Natural Multimodal Dialogue*, Verona, Italy.

Stiefelhagen, R., Yang, J. 1997. Gaze tracking for multimodal human-computer interaction. *Proc. Int. Conf. on Acoustics, Speech and Signal Processing*.

Stolcke, A., Konig, Y., Weintraub, M. 1997. Explicit word error minimization in N-best list rescoring. *Proc. Eurospeech*, pp. 163–166.

Streit, M., Batliner, A., Portele, T. 2004. Cognitive-model-based interpolation of emotions in a multi-modal dialogue system. *Proc. Affective Dialogue Systems: Tutorial and Research Workshop*, pp. 65–76.

Sturm, J., Bakx, I., Cranen, B., Terken, J., Wang, F. 2002. Usability evaluation of a Dutch multimodal system for train timetable information. *Proc. Language Resources and Evaluation Conference*, Las Palmas, Canary Islands, Spain.

Su, M., Huang, H., Lin, C., Huang, C., Lin, C. 1998. Application of neural networks to spatio-temporal hand gesture recognition. *Proc. IEEE World Congress on Computational Intelligence*, USA, pp. 2116–2121.

Su, Q., Silsbee, P. 1996. Robust audiovisual integration using semicontinous Hidden Markov Models. *Proc. 4th International Conference on Spoken Language Processing*, pp. 42–45.

Su, Y., Zheng, F., Huang, Y. 2001. Design of a semantic parser with support to ellipsis resolution in a Chinese spoken dialogue system. *Proc. Eurospeech*, pp. 2161–2164.

Sumi, Y., Ohta, Y. 1995. Detection of face orientation and facial components using distributed appearance e modelling. *Proc. First International Workshop on Automatic Face and Gesture Recognition*, pp. 254–259.

Takeuchi, A., Nagao, K. 1995. Situated facial displays: Towards social interaction. *Proc. SIGCHI*, pp. 450–454.

Taylor, M. M., Néel, F., Bouwhuis, D. (eds) 2000. *The Structure of Multimodal Dialogue II*. John Benjamins Publishing.

Terken, J., Te Riele, S. T. 2001. Supporting the construction of a user model in speech-only interfaces by adding multi-modality. *Proc. Eurospeech*, pp. 2177–2180.

Terzopoulos, D., Waters, K. 1990. Physically-based facial modelling, analysis, and animation. *Journal of Visualization and Computer Animation*, vol. 1, no. (2), pp. 73–90.

Torres, F., Sanchís, E., Segarra, E. 2003. Development of a stochastic dialog manager driven by semantics. *Proc. Eurospeech*, pp. 605–608.

Traum, D., Rickel, J. 2001. Embodied agents for multi-party dialogue in immersive virtual worlds. *Proc. Workshop on Representing, Annotating and Evaluating Non-verbal and Verbal Communicative Acts to Achieve Contextual Embodied Agents*, pp. 53–58.

Trower, T. 2000. *Microsoft Agent*. http://www.microsoft.com/msagent. Microsoft Corporation, March 2000.

Tsang-Long, P., Yu-Te Chen, J.-J. L., Jun-Heng, Y. 2004. The construction and testing of a Mandarin emotional speech database. *Proc. XVI Conference on Computational Linguistics and Speech Processing*, Taipei, Taiwan.

Tsutsui, T., Saeyor, S., Ishizuka, M. 2000. MPML: A multimodal presentation language with character control functions. *Proc. World Conference on the WWW and Internet* (WebNet 2000), San Antonio, pp. 537–543.

Turunen, M., Hakulinen, J. 2000. Jaspis – A framework for multilingual adaptive speech applications. *Proc. of International Conference on Spoken Language Processing* (ICSLP), pp. 719–722.

Turunen, M., Hakulinen, J. 2003. Jaspis2 – An architecture for supporting distributed spoken dialogues. *Proc. Eurospeech 2003*, pp. 1913–1916.

Turunen, M., Salonen, E.-P., Hartikainen, M., Hakulinen, J., Black, W., Ramsay, A., Funk, A., Conroy, A., Thompson, P., Stairmand, M., Jokinen, K., Rissanen, J., Kanto, K., Kerminen, A., Gamback, B., Cheadle, M., Olsson, F., Sahlgren, M. 2004a. AthosMail - a Multilingual Adaptive Spoken Dialogue System for E-mail Domain. In *Proceedings of Workshop on Robust and Adaptive Information Processing for Mobile Speech Interfaces*, Geneva, pp. 77–86.

Turunen, M., Salonen, E., Hartikainen, M., Hakulinen. J. 2004b. Robust and adaptive architecture for multilingual spoken dialogue systems. *Proc. International Conference on Spoken Language Processing* (ICSLP), pp. 3081–3084.

Van de Burgt, S. P., Andernach, T., Kloosterman, H., Bos, R., Nijholt, A. 1996. Building dialogue systems that sell. *Proc. NLP and Industrial Applications*, pp. 41–16.

Van Mulken, S., André, E., Müller, J. 1998. The Persona effect: How substantial is it?. *Proc. Human Computer Interaction*, pp. 53–66.

Vilar, D., Castro, M. J., Sanchís, E. 2003. Connectionist classification and specific stochastic models in the understanding process of a dialogue system. *Proc. Eurospeech*, pp. 645–648.

Vo, M., Waibel, A. 1997. Modeling and interpreting multimodal inputs: A semantic interpretation approach. *Technical report CMU-CS-97-192*, Carnegie Mellon University.

Vo, M. T., Wood, C. 1996. Building and application framework for speech and pen input integration in multimodal learning interfaces. *Proc. International Conference on Acoustics, Speech and Signal Processing*, Atlanta, GA, pp. 3545–3548.

Wachs, J., Kartoun, U., Stern, H., Edan, Y. 2002. Real-time hand gesture telerobotic system using fuzzy C-Means clustering. *Proc. 12th Annual Conference of Industrial Engineering and Management*, pp. 403–409.

Wahlster, W. 1993. Verbmobil, Translation of face-to-face dialogs. *Proc. Eurospeech*, pp. 29–38.

Wahlster, W. 1994. Overview of user modelling for Verbmovil. Keynote presentation to User Modeling biannual workshop, Vienna, June.

Wahlster, W. 2002. SmartKom: Fusion and fission of speech, gestures, and facial expressions. *Proc. First International Workshop on Man-Machine Symbiotic Systems*, pp. 213–225.

Wahlster, W., Reithinger, N., Blocher, A. 2001. SmartKom: Multimodal communication with a life-like character. *Proc. Eurospeech*, pp. 1547–1550.

Waibel, A., Jain, A., McNair, A., Saito, H., Hauptmann, A., Tebelkis, J. 1991. JANUS: A speech-to-speech translation system using connectionist and symbolic processing strategies. *Proc. Int. Conf. on Acoustics, Speech and Signal Processing*, pp. 793–796.

Waibel, A., Vo., M. T., Duchnowski, P., Manke, S. 1996. Multimodal interfaces. *Artificial Intelligence Review*, vol. 10, no. (3–4), pp. 299–316.

Walker, J. H., Sproull, L., Subramami, R. 1994. Using a human face in an interface. *Proc. SIGCHI Conference on Human Factors in Computing Systems*, pp. 85–91.

Walker, M. A. 1994a. Discourse and deliberation: testing a collaborative strategy. *Proc. 15th International Conference on Computational Linguistics: COLING 94*, pp. 1205–1211.

Walker, M. A. 1994b. Experimentally evaluating communicative strategies: The effect of the task. *Proc. Conference of the American Association Artificial Intelligence, AAAI94*, pp. 86–93.

Walker, M. A., Litman, D., Kamm, C. A., Abella, A. 1997. PARADISE: a General framework for evaluating spoken dialogue agents. *Meeting of the ACL*, pp. 271–280.

Walker, M. A., Passonneau, R., Boland, J. E. 2001. Quantitative and qualitative evaluation of DARPA Communicator spoken dialogue systems. *Proc. of the Meeting of the ACL, ACL 2001*, pp. 515–522.

Walker, M. A., Rudnicky, A., Rashmi, P., Aberdeen, J., Owen, E., Garafolo, J., Hastie, H., Le, A., Pellom, B., Potamianos, A., Passonneau, R., Roukos, S., Sanders, G., Seneff, S., Stallard, D. 2002. DARPA Communicator: Cross-system results for the 2001 evaluation. *Proc. International Conference on Spoken Language Processing (ICSLP)*, pp. 269–272.

Wasinger, R., Stahl, C., Krueger, A. 2003. Robust speech interpretation in a mobile environment through the use of multiple and different media input types. *Proc. Eurospeech*, pp. 1049–1052.

White, M. 2004. Reining in CCG chart realization. *Proc. 3rd International Conference on Natural Language Generation*, pp. 182–191.

Whittaker, S., O'Conaill, B. 1997. The role of vision in face-to-face mediate communication. K. Finn, A. Sellen, S. Wilbur (eds). *Video Mediated Communication*, Lawrence Erlbaum Associates, Inc., pp. 23–49.

Williams, G., Renals, S. 1997. Confidence Measures for Hybrid HMM/ANN speech recognition. *Proc. Eurospeech*, pp. 1955–1958.

Wittenburg, P., Broeder, D., Offenga, F., Willems, D. 2002b. Metadata set and tools for multimedia/multimodal language resources. *Proc. 3rd Multimodal Resources and Multimodal Systems Evaluation Workshop*, Las Palmas, Canary Islands, Spain, pp. 9–14.

Wittenburg, P., Kita, S., Brugman, H. 2002a. Cross-linguistic studies of multimodal communication. *Proc. Multimodal Resources and Multimodal Systems Evaluation Workshop*, Las Palmas, Canary Islands, Spain, pp. 25–32.

Woltjer R., Tan, W. J., Chen, F. 2003. Speech-based, manual-visual, and multi-modal interaction with an in-car computer – evaluation of a pilot study. *Proc. Eurospeech*, pp. 1053–1056.

Wu, L., Oviatt, S., Cohen, P. 1999. Multimodal integration: A statistical view. *IEEE Transactions on Multimedia*, vol. 1, no. (4), pp. 334–342.

Wu, L., Oviatt, S. L., Cohen, P. R. 2002. From members to teams to committee – A robust approach to gestural and multimodal recognition. *IEEE Transactions on Neural Networks*, vol. 13, no. (4), pp. 972–982.

Xiao, B., Girand, C., Oviatt, S. L. 2002. Multimodal integration patterns in children. *Proc. International Conference on Spoken Language Processing (ICSLP)*, pp. 629–632.

Xu, Y., Araki, M., Niimi, Y. 2003a. A mutilingual-supporting dialog system across multiple domains, *Journal of Acoustical Science and Technology*, vol. 24, no. (6), pp. 349–357.

Xu, Y., Di, F., Araki, M., Niimi, Y. 2003b. Methods to improve its portability of a spoken dialog system both on task domains and languages, *Proc. Eurospeech 2003*, pp. 1901–1904.

Yambor, W., Draper, B., Beveridge, J. 2000. Analyzing PCA-based face recognition algorithms: Eigenvector selection and distance measures. *Proc. Second Workshop on Empirical Evaluation Methods in Computer Vision*, Dublin, Ireland, 1 July 2000.

Yang, J., Stiefelhagen, R., Meier, U., Waibel, A. 1998. Real-time face and facial feature tracking and applications. *Proc. Workshop on Audio-Visual Speech Processing*, pp. 79–84.

Yang, J., Yang, W., Denecke, M., Waibel, A. 1999. Smart sight: A tourist assistant system. *Proc. IEEE Third International Symposium on Wearable Computers*, pp. 73–78.

Yang, M.-H., Kriegman, J., Ahuja, N. 2002. Detecting faces in images: A survey. *IEEE Transactions on Pattern Analysis and Machine Intelligence*, vol. 24, no. (1), pp. 34–58.

Yasuda, H., Takahashi, K., Matsumoto, T. 2000. A discrete HMM for online handwriting recognition. *Pattern Recognition and Artificial Intelligence*, vol. 14, no. (5), pp. 675–689.

Young, S., Adda-Decker, M., Aubert, X., Dugast, C., Gauvain, J. L., Kershaw, D. J., Lamel, L., Leeuwen, D. A., Pye, D., Robinson, A. J., Steeneken, H. J. M., Woodland, P. C. 1997. Multilingual large vocabulary speech recognition: the European SQUALE project. *Computer Speech and Language*, vol. 11, pp. 73–89.

Zitouni, I., Siohan, O., Lee, C. H. 2003. Hierarchical class n-gram language models: Toward better estimation of unseen events in speech recognition. *Proc. Eurospeech*, pp. 237–240.

Zoltan-Ford, E. 1991. How to get people to say and type what computers can understand. *Man-Machine Studies*, vol. 34, no. (4), pp. 517–547.

Zue, V., Seneff, S., Glass, J., Polifroni, J., Pao, C., Hazen, T., Hetherington, L. 2000. Jupiter: A telephone-based conversational interface for weather information. *IEEE Trans. on Speech and Audio Proc.*, vol. 8, no. (1), pp. 85–96.

Index

Abort   132
Acoustic model   20, 21, 152
Action
  class   110, 111, 112
  form bean   110, 111, 112, 113, 115
  planner   6, 136
Activation mode
  automatic   18
  barge-in   18, 19
  hotword   18
  manual   18
  Push & Talk   18, 46, 178
AdApt dialogue system   8, 45, 46, 52, 59, 61,
  62, 134
Agent
  animated   6, 44, 70, 85, 212
Agreeableness   82, 83, 84, 141
Algorithm
  Forward   21
  Forward-Backward   21
  Viterbi   21, 154
Ambiguity
  lexical   23
  structural   23
  syntactic   23
Amitiés dialogue system   42
Anaphora   2, 23, 47, 52, 95, 97, 137
Animated agent   6, 44, 70, 85, 210
Annotation
  gesture   14, 122
  morphosyntactic   14, 118, 149

multimodal corpora   15, 122, 185, 186
pragmatic   14, 118, 119, 120,
  123, 149
prosodic   14, 118, 119
speech   14, 122, 123
spoken corpora   14, 118, 149
syntactic   14, 118
Anticipation/retention   65
Anvil annotation system   15, 85, 186
APML   78, 179
Architecture
  dialogue-control centred   86, 94, 108, 141
  Galaxy Communicator   67
  hub-and-spoke   98, 99
  interlingua   91, 92, 93, 117, 142
  OAA   68, 69
  semantic-frame conversion   86, 92, 93
  system   5, 7, 69, 84, 90, 112, 117, 146,
  147, 148
Arm   6, 30, 31, 123, 180
ARPA   189
ASP   163, 172
ASR
  acoustic confusion   19
  acoustic variability   18
  coarticulation   19
  environmental conditions   19, 80
  error
    deletion   19, 24
    false alarm   19
    insertion   19, 24